The Horizons of Evolutionary Robotics

Intelligent Robots and Autonomous Agents

Edited by Ronald C. Arkin

For a complete list of the books published in this series, please see the back of this book.

The Horizons of Evolutionary Robotics

Edited by Patricia A. Vargas, Ezequiel A. Di Paolo, Inman Harvey, and Phil Husbands

The MIT Press
Cambridge, Massachusetts
London, England

MIT Press books may be purchased at special quantity discounts for business or sales promotional use. For information, please email special_sales@mitpress.mit.edu or write to Special Sales Department, The MIT Press, 55 Hayward Street, Cambridge, MA 02142.

This book was set in ITC Stone Serif Std 9/13 pt by Toppan Best-set Premedia Limited, Hong Kong. Printed and bound in the United States of America.

Library of Congress Cataloging-in-Publication Data
The horizons of evolutionary robotics / edited by Patricia A. Vargas, Ezequiel A. Di Paolo, Inman Harvey, and Phil Husbands.
   pages cm.—(Intelligent robotics and autonomous agents)
Includes bibliographical references and index.
ISBN 978-0-262-02676-5 (hardcover : alk. paper)
1. Evolutionary robotics. I. Vargas, Patricia A., 1969– II. Di Paolo, Ezequiel A. III. Harvey, Inman. IV. Husbands, Phil.
TJ211.37.H65 2014
629.8'92—dc23
2013025304

10 9 8 7 6 5 4 3 2 1

*To my parents, Joaquim Amâncio Filho and Diná Vargas Amâncio—*P.A.V.

Contents

Preface

Evolutionary robotics (ER) is a novel field of research that aims to apply evolutionary computation techniques to evolve the overall design or controllers, or both, for real and simulated autonomous robots. The origins of ER date back to the beginning of the 1990s. Since then it has been developed through various research centers around the world and it is still attracting the attention of an increasing number of researchers. ER now has many research groups worldwide, generally fully devoted to uncovering the intricacies of this promising multidisciplinary area of research.

Since ER's birth, there have been a number of international conferences and workshops devoted to the field, as well as numerous sessions at more general conferences (e.g., the From Animals to Animats series, Artificial Life series of conferences, the European Conference on Artificial Life [ECAL series], and special sessions at the Genetic and Evolutionary Computation Conference [GECCO] and IEEE Congress on Evolutionary Computation [CEC] series of conferences). As a result some conference proceedings are available. To date, however, the only textbook was published in 2000 by Stefano Nolfi and Dario Floreano (*Evolutionary Robotics: The Biology, Intelligence, and Technology of Self-Organizing Machines*, MIT Press). Hence after many years of further research, and many hundreds of published papers, we believe it is time to produce a new and authoritative overview of the field. This book not only revisits the most important work in the area, but also includes novel investigations, emerging discoveries, and cutting-edge developments.

The main purpose of this volume is to present a lively, extensive compilation of original articles on the cross-fertilization between ER and other research areas. Bearing this in mind, the book seeks to provide complete coverage of the foremost achievements of the field to date. This was accomplished by inviting leading practitioners to contribute chapters written in an appealing style aimed at engaging a wide academic readership. The contributors include neuroscientists, cognitive scientists, philosophers, engineers, computer scientists, and robotics engineers.

The Horizons of Evolutionary Robotics is intended to be a reference and a guide for future advances, not only highlighting the primary predictions and suggestions from today's leading researchers, but also showing how evolutionary robotics is helping us to shape a new kind of interdisciplinary science.

1 Context and Challenges for Evolutionary Robotics

Patricia A. Vargas, Ezequiel A. Di Paolo, Inman Harvey, and Phil Husbands

1.1 Context

Evolutionary robotics involves the use of evolutionary computing techniques to automatically develop some or all of the following properties of a robot: the control system, the body morphology, sensor and motor properties, and layout. Basically, populations of artificial genomes encode properties of autonomous (usually mobile) robots required to carry out a particular task or to exhibit some set of behaviors. The genomes are mutated and interbred creating new generations of robots according to a Darwinian scheme in which the fittest individuals are most likely to produce offspring. Fitness is measured in terms of how good a robot's behavior is according to some evaluation criteria. This is usually automatically measured but may be also based on the experimenter's judgment.

The origins of ER as we know it today date back to the beginning of the 1990s (Husbands and Harvey 1992; Beer and Gallagher 1992; Cliff, Harvey, and Husbands 1993; Parisi and Nolfi 1993; Floreano and Mondada 1994). Since then it has been developed through various research centers around the world, where one can find groups of scientists devoted to uncovering the intricacies of this promising multidisciplinary area of research.

Due to its interdisciplinary character, ER is being employed in the development of other fields of research, including neuroscience and cognitive science. We envisage that ER will play a major role in the future in a variety of industrial applications, including in the entertainment industries, exploration and navigation applications, and technology relating to environmental issues, not to mention civilization itself with the advent of robot companions, for which autonomy and robustness are fundamental requirements.

The aims of this chapter are to briefly describe ER's main techniques and methods, and to place the other chapters in this book in the wider context of the field.

1.2 The Basic Evolutionary Robotics Methodology

This section introduces the ER methodology and can be skipped by those familiar with the field. It is provided so that those who are not acquainted with the methods used can readily grasp the techniques referred to and terminology used throughout the book. For a detailed review of ER applications, and a more technical coverage of techniques, see Floreano, Husbands, and Nolfi (2008).

Alan Turing's (1950) article "Computing Machinery and Intelligence" is widely regarded as one of the seminal works in artificial intelligence. It is best known for what came to be called the Turing test—a proposal for deciding whether or not a machine is intelligent. However, tucked away toward the end of Turing's wide-ranging discussion of issues arising from the test is a far more interesting proposal. He suggests that worthwhile intelligent machines should be adaptive, should learn and develop, but concedes that designing, building, and programming such machines by hand is probably infeasible. He goes on to sketch an alternative way of creating machines based on an artificial analog of biological evolution. Each machine would have hereditary material encoding its structure, mutated copies of which would form offspring machines. A selection mechanism would be used to favor better-adapted machines—in this case those that learned to behave most intelligently. Turing proposed that the selection mechanism should largely consist of the experimenter's judgment.

It was more than forty years before Turing's long forgotten suggestions became reality. In the early 1990s, emerging out of the spirited milieu of "New AI" (Brooks 1999) and building on the development of principled evolutionary search algorithms (Holland 1975), researchers at the National Research Council (CNR), Rome; Ecole Polytechnique Federale de Lausanne (EPFL); the University of Sussex; and Case Western University independently demonstrated methodologies and practical techniques to evolve, rather than design, control systems for primitive intelligent machines (Cliff, Harvey, and Husbands 1993; Beer and Gallagher 1992; Parisi and Nolfi 1993; Floreano and Mondada 1994). Initial motivations were similar to Turing's: the hand design of intelligent adaptive machines intended for operation in natural environments is extremely difficult. Would it be possible to wholly or partly automate the process?

The scheme used in most ER research is some elaboration or other of the basic methodology illustrated in figure 1.1. The key elements of the evolutionary robotics approach are:

• A method for measuring the fitness of the robot behaviors generated from these genomes.

• A way of applying selection and a set of "genetic" operators to produce the next generation from the current one.

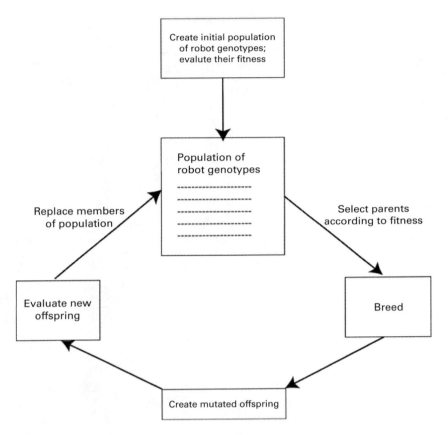

Figure 1.1
General scheme employed in evolutionary robotics.

The general scheme is like that of any application of an evolutionary search algorithm. However, many details of specific parts of the process, particularly the evaluation step, are peculiar to evolutionary robotics.

Genetic Encoding

While many aspects of the robot design can potentially be under genetic control, *at least* the control system always is. By far the most popular form of controller used in ER is some sort of neural network. These range from straightforward feedforward networks of simple elements (Floreano and Mondada 1994) to relatively complex dynamic and plastic recurrent networks (Beer and Gallagher 1992; Floreano and Urzelai 2000; Philippides et al. 2005), as illustrated in figure 1.2. In the simplest case, a fixed architecture network is used to control a fixed robot whose sensors feed into the network

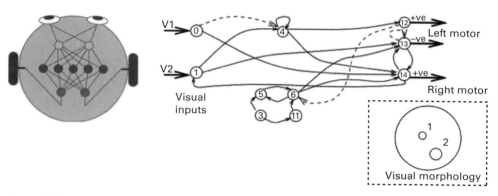

Figure 1.2
Evolved neurocontrollers. On the left a simple fixed architecture feedforward network is illustrated. The connection weights, and sometimes the neuron properties, are put under evolutionary control. On the right a more complex architecture is illustrated. In this case the whole architecture, including the number of neurons and connections, is under evolutionary control, along with connection and neuron properties and the morphology of a visual sensor that feeds into the network (Cliff, Harvey, and Husbands 1993). The inputs V1 and V2 refer to signals derived from genetically specified "patches" of a camera input, as shown on the inset on the right of the figure.

that in turn feeds out to the robot motors. In this scenario the parameters of the network (connection weights and relevant properties of the units such as thresholds or biases) are coded as a fixed length string of numerical values.

A more complex case, which has been explored since the very early days of evolutionary robotics (Cliff, Harvey, and Husbands 1993), involves the evolution of the network architecture as well as the properties of the connections and units. Typically the size of the network (number of units and connections) and its architecture (wiring diagram) are unconstrained and free to evolve. This involves more complex encodings that can grow and shrink, as units and connections are added or lost, while allowing a coherent decoding of connections between units. These range from relatively simple strings employing blocks of symbols that encode a unit's properties and connections relative to other units (Cliff, Harvey, and Husbands 1993) to more indirect schemes that make use of growth processes in some geometric space (Philippides et al. 2005) or employ genetic programming like tree representations in which whole subbranches can be added, deleted, or swapped over (Gruau 1995).

The most general case involves the encoding of control network and body and sensor properties. Various kinds of developmental schemes have been used to encode the construction of body morphologies from basic building blocks, both in simulation and in the real world (Sims 1994). Here the inspiration is developmental biology and the genetic encoding is not simply a string of variable values that directly map onto robot properties, but a set of parameters that control an indirect growth process that results in a robot or robot controller.

The details of the specific evolutionary algorithm used to manipulate the genetic encoding vary widely. Many find rather simple genetic algorithms (GAs) sufficient (Floreano and Mondada 1994); others have found more sophisticated GA variants, involving geographical distribution and multiple levels of operators, particularly useful for complex fitness landscapes (Philippides et al. 2005); still others have developed interesting new types of evolutionary algorithms specifically aimed at evolving neurocontrollers (Stanley and Miikkulainen 2002).

Fitness Evaluation

Fitness evaluations consist of translating the genome in question into a robot instantiation and then measuring aspects of the resulting behavior. In the earliest work aimed at using evolutionary techniques to develop neurocontrollers for particular physical robots, members of a population were downloaded in turn onto the robot and their behavior was monitored and measured either automatically by clever experimental setups (Floreano and Mondada 1994; Harvey, Husbands, and Cliff 1994) or manually by an observer (Gruau and Quatramaran 1997). The machinery of the evolutionary search algorithm was managed on a host computer while the fitness evaluations were undertaken on the target robot.

One drawback of evaluating fitness on the physical robot is that this cannot be done any quicker than real time, making the whole evolutionary process rather slow. In the early work in the field this approach was taken because researchers considered it unlikely that simulations could be made accurate enough to allow good transfer of evolved behavior onto the real robot. However, a careful study of accurate physics-based simulations of a Khepera robot, with various degrees of noise added, proved this assumption false (Jakobi, Husbands, and Harvey 1995). This led to various successful simulation-based approaches including Jakobi's minimal simulation methodology (Jakobi 1998) whereby computationally very efficient simulations are built by modeling only those aspects of the robot-environment interaction deemed important to the desired behavior and masking everything else with carefully structured noise (so that evolution could not come to rely on any of those aspects). An alternative approach uses plastic controllers that further adapt through self-organization to help smooth out the differences between an inaccurate simulation and the real world (Urzelai and Floreano 2001). Instead of evolving connection weights, in this approach "learning rules" for adapting connection strengths are evolved—this results in controllers that continually adapt to changes in their environment. For details of further approaches, see Floreano, Husbands, and Nolfi (2008). Much evolutionary robotics work now makes use of simulations; without them it would be impossible to do the most ambitious work on the concurrent evolution of controllers and body morphology (Lipson and Pollack 2000). However, although simulation packages and techniques have developed rapidly in the past few years, there will still inevitably be discrepancies between simulation and reality and the lessons and insights of the work outlined in this chapter should not be forgotten.

An interesting distinction can be made between implicit and explicit fitness functions in evolutionary robotics (Nolfi and Floreano 2000). In this context, an explicit fitness function rewards specific behavioral elements—such as traveling in a straight line—and hence shapes the overall behavior from a set of specific behavioral primitives. Implicit fitness functions operate at a more indirect, abstract level—fitness points are given for completing some task but they are not tied to specific behavioral elements. Implicit fitness functions might involve components such as maintaining energy levels, covering as much ground as possible, or just remaining "viable" in some sense—all components that can be achieved in many different ways. In practice it is quite possible to define a fitness function that has both explicit and implicit elements. In some ER work, crafting a suitable fitness function has been found to be crucial in producing successful outcomes (e.g., Quinn et al. 2003). In some cases the evolutionary process proceeds in discrete stages where competences are gradually built up in an incremental way (often through additions to the current solution); here the fitness function may well change as evolution progresses (Harvey, Husbands, and Cliff 1994).

Advantages

Potential advantages of the ER methodology include:

• The ability to explore potentially unconstrained designs that have large numbers of free variables. A *class* of robot systems (to be searched) is defined rather than specific fully defined robot designs. This means fewer assumptions and constraints are necessary in specifying a viable solution.
• The ability to use the methodology to fine tune parameters of an already successful design, or to build on an existing solution.
• The ability, through the careful design of fitness criteria and selection techniques, to take into account multiple, and potentially conflicting, design criteria and constraints (e.g., efficiency, cost, weight, power consumption).
• The possibility of developing highly unconventional and minimal designs.
• The ability to explicitly take into account robustness and reliability as major driving forces behind the fitness measure, factors that are particularly important for certain applications.

The first of these advantages has made ER an attractive tool for exploring specific scientific questions. Because it allows a relaxation of assumptions about mechanisms underlying the generation of behavior, as well as the synthesis of behaviors under alternative sets of assumptions, it is an interesting way of approaching fundamental questions in neuroscience and cognitive science. ER specifies the target behavior but doesn't specify how it should be achieved. Various examples of this scientific use will be found throughout this book.

1.3   The Evolutionary Robotics Chronicle

From the earliest work at the start of the 1990s, most evolved robot controllers were, and indeed still are, based on artificial neural networks (Husbands and Harvey 1992; Beer and Gallagher 1992; Harvey, Husbands, and Cliff 1994; Parisi and Nolfi 1993; Floreano and Mondada 1994). This was partly because ER was built on a huge surge of interest in biologically inspired methods, including neural networks, which had been sidelined for many years, and partly because neural net's inherently cellular nature was deemed to be suited to evolution (i.e., highly evolvable) by early practitioners (Husbands and Harvey 1992; Cliff, Harvey, and Husbands 1993). Even in those pioneering days, some of the neural models and architectures used were directly inspired by contemporary neuroscientific findings. As more and more subtleties of the nervous system have been slowly revealed by major advances in neuroscience over the past few decades, it has become clear that the complex mechanisms underlying natural behavior are far removed from the simple connectionist models that until recently dominated the study of neural networks. Neuroscience can thus provide a rich vein of inspiration for the development of more sophisticated neural models that can be exploited by ER. But given the complexity of the systems, many of the details of the mechanisms at play in biological nervous systems are not well understand, which gives ER a potential role in exploring basic scientific questions in neuroscience by synthesizing model systems making use of, or indeed ignoring, certain assumptions about mechanisms. Phil Husbands, Renan C. Moioli, Yoonsik Shim, Andy Philippides, Patricia A. Vargas, and Michael O'Shea explore these issues in chapter 2 where they discuss the links between ER and neuroscience. In their contribution, the authors show how principles from neuroscience, including volume signaling, chaotic dynamics, and phase coupling, can be profitably incorporated into ER methods, and how questions in neuroscience and cognitive science, such as the role of neural synchronization, can be explored using the ER methodology.

At its inception ER was part of a wider movement opposed to the prevailing orthodoxy in AI (artificial intelligence) and cognitive science. One of the strands of work in this insurgency was a newfound interest in dynamical approaches to cognition. Whereas the mainstream view involved a one-step-at-a-time model based around a sense-think-act consecutive pipeline of (computational, representational) processing, the dynamical approach threw out the idea of centrally clocked computational models. Instead, behavior generation was viewed in terms of dynamical interactions between agent and environment (or more precisely brain-body-environment interactions) (Port and van Gelder 1995), in a way that harked back to the earlier cybernetics frameworks developed by Ashby (1952). The agent's nervous system was thought of as a free-running dynamical system, operating in a distributed, most likely asynchronous, way with many different timescales at play. Hence, with the advent of newly evolved intelligent robot controllers it is not surprising that the quest for understanding their evolved dynamics began immediately (Beer 1995; Husbands, Harvey, and Cliff 1995; Tani and Nolfi 1999) and has continued ever since (Beer 2003; Negrello and Pasemann 2008; Williams, Beer, and Gasser 2008a; Izquierdo, Harvey, and Beer 2008; Izquierdo and Buhrmann 2008). Some of these projects were early examples of the use of ER to synthesize models that did not adhere to the mainstream assumptions, in this case dynamical rather than computational behavior-generating systems. By synthesizing and then analyzing, existence proofs of alternative, dynamical mechanisms were produced. In chapter 3, Randall D. Beer, one of the pioneers of these approaches, provides an insightful step-by-step guide to the dynamical analysis of evolved agents. He helps to demystify the sometimes dark and difficult arts of dynamical analysis.

Applying evolution to create robot behaviors is often a far from straightforward task and sometimes ER researchers find themselves adrift in a sea of confusion when faced with the numerous choices to be made in setting parameters and designing experiments. There are many factors that shape the search space and determine which evolutionary pathways are followed, but not all pathways are equally fruitful. With a view to shedding some light on the challenging design tasks faced by ER researchers, in chapter 4 Inman Harvey and Ezequiel A. Di Paolo underline "the various features of evolutionary paths that can make them more or less easy for evolution to navigate."

Along with a discussion of the mechanisms of evolutionary search they offer various heuristics (or "travel tips") to help with the construction of search spaces replete with evolutionary pathways that are likely to lead to the desired goal in an efficient way.

This might involve embracing, rather than being scared of, such search space properties as neutrality (Barnett 2001), or being aware of the intricacies of some of the building blocks being used, such as the style of neural network, so that parameter ranges can be set in a helpful way, or considering an incremental approach (Harvey, Husbands, and Cliff 1994). Many of the heuristics thus presented are part of the folklore but have never before been properly committed to print; hence this chapter provides a helpful initial guide to point explorers in the right direction.

Navigation and exploration are among the most studied behaviors in ER, and early on researchers raised an important question: do the evolved robots require an internal map-like representation of the environment? (Nolfi et al. 1994; Miglino, Lund, and Nolfi 1995; Miglino, Nafasi, and Taylor 1996). More generally, what kinds of mechanisms underlie spatial orientation abilities? In other words: how does spatial cognition work? Orazio Miglino and Michela Ponticorvo address this issue in chapter 5 through a detailed discussion of an extensive program of work that uses the ER approach to explore spatial cognition. In particular they use the methodology to synthesize models that allow them to discard orthodox views of spatial cognition that assume internal maps and geometric representations of the world. The chapter is appealingly structured around the gradual development of an artificial agent whose competences increase by reference to various experiments carried out by Miglino and colleagues.

In nearly all early ER work the aim was to evolve control systems for preexisting robots: the brain was constrained to fit a particular body and set of sensors. Of course in nature the nervous system evolved simultaneously with the rest of the organism. As a result, the nervous system is highly integrated with the sensory apparatus and the rest of the body: the whole operates in a harmonious and balanced way—there are no distinct boundaries between control system, sensors, and body. However, even from the earliest days there was an acknowledgment that the question of the degree to which body morphology can influence an agent's behavior is important in the quest for more complex agents. Karl Sim's inspirational work on simulated coevolving creatures, whose bodies as well as brains were under evolutionary control, was the first significant investigation in this direction (Sims 1994). There were severe technical difficulties in translating such investigations into the realm of real physical robots until Lipson and Pollack's groundbreaking work on the Golem project (Lipson and Pollack 2000). In that project they pushed the idea of fully evolvable robot hardware about as far as was reasonably technologically feasible at the time. Autonomous "creatures" were evolved in simulation out of basic building blocks (neurons, bars, actuators). The fittest individuals were then fabricated robotically using rapid manufacturing technology (plastic extrusion 3D printing). The team thus achieved autonomy of design and construction using evolution in a "limited universe" physical simulation coupled to automatic fabrication. More recently investigations in this area have started to multiply (Bongard 2011; Clark et al. 2012; Long 2012). In chapter 6, Josh Bongard elegantly reflects on the potential advantages of and issues involved in simultaneously evolving robot brains and bodies. He explores problems central to the balanced interplay among body morphology, neural processing, and environmental interactions in the generation of embodied adaptive behavior, and develops a set of design principles for intelligent systems in which these issues take center stage.

Another kind of complexity arises when dealing with more than one robot. The domains of social interactions, collective behavior, and communication have provided challenging goals for evolutionary robotics and autonomous robotics in general since very early on. In the 1950s, Grey Walter added a lamp to two of his famous light-seeking tortoises that would switch on or off depending on their current behavior (Walter 1953). The result was that the two robots would be intermittently attracted to each other and enact a sophisticated dance. From these complex interaction patterns it would have been nearly impossible to reverse engineer the actual sensorimotor control programmed by Walter, which was in fact relatively simple. Work in evolutionary robotics has been influenced by such inspirations, in particular following an underlying intuition that the social domain is a kind of amplifier of the sophistication, intelligence, and complexity of the evolved behaviors in mobile robots (Di Paolo 2000; Quinn 2001; Cangelosi and Riga 2006, Williams, Beer, and Gasser 2008b). Collective tasks demand the solution of specific problems, such as how should robots coordinate their behavior, but the intuition is that the gains in the overall performance and competence of the robots are likely to outweigh these additional difficulties because collective intelligence is widespread in nature. In the case of humans, higher mental functions seem closely connected with our social nature, and it may be that certain kinds of higher cognitive abilities, in both animals and machines, can only be developed in a social context.

An interesting example of evolved group behavior is self-organized coordinated movement without an explicit leader, as pioneered by Quinn et al. (2003). In Quinn and coauthors' seminal work a group of identical robots, equipped only with a small numbers of short-range infrared (IR) sensors, were evolved to move in close formation starting from initial random orientations. This required a direction of movement and suitable individual robot orientations to emerge from group dynamics: what could be described as the emergence of a "leader" and "followers." In chapter 7, Vito Trianni, Elio Tuci, Christos Ampatzis, and Marco Dorigo explore group behavior using a similar scenario, but with an added twist or two. They use a group of identical robots (s-bots), which can form physical connections to self-assembly into a larger robotic entity built from individual s-bots. They demonstrate that evolutionary robotics techniques can be used to develop controllers that allow the s-bots to perform coordinated movement in quite challenging environments in which holes have to be avoided (or at least bridged, as the s-bots can form rigid structures that enable them to support each other). A notable aspect of this work is the way in which the controllers exploit physical interactions between the robots to support emergent self-organization of the group. The chapter extensively develops the theme of group behavior by reporting results from successful experiments in which the s-bots are evolved to perform synchronization behaviors, collective decision making, and behaviors requiring automatic allocation of different roles. An important lesson learned from this research is the importance to evolvability of the right balance between controller complexity and possible modes of interaction among the robots, as defined by the hardware.

Even if a cooperative context is already assumed, several open questions remain in the evolution of pro-social behavior that evolutionary robotics experiments could shed some light on. For instance, one possible such area is the origins of communicative signals. In chapter 8, Joachim de Greeff and Stefano Nolfi neatly attempt to answer some of these questions by contrasting existing theories about the origins of animal communication with experiments in simulated and real robotic scenarios requiring some form of behavioral coordination between two robots. Their "synthetic experiments" make concrete ideas such as the origin of signal production and signal understanding and their relation to other, nonsignaling behavior. The robots must achieve a coordination task, moving between different areas in an arena, occupying them at the same time and then switching positions repeatedly. The task requires several layers of coordination (Where to move first? How long to stay there? Is the other robot also in its place? At what point shall we switch?). The setup permits the presence of implicit (body movements) signals and explicit (via a dedicated communication channel) signals. Successful performance seems to require some level of explicit communication (as evidenced by the failure to replicate such performance levels when the dedicated signaling channels are knocked down). All signals observed seem to be grounded in the current sensorimotor context of the signaler (they are deictic). The robots also use implicit forms of bodily coordination, demonstrating a spectrum of possible communicative layers of different complexity that may have impact on the understanding of how signals acquire their communicative functionality in the first place.

ER work on social and group behavior is both inspired by biology and can at the same time illuminate current theories and debates in biology. The contribution by Sabine Hauert, Sara Mitri, Laurent Keller, and Dario Floreano in chapter 9 is an example of this two-way street. By setting themselves the task of better understanding the evolution of cooperation in teams of robots, the authors superbly draw inspiration from evolutionary theory to fine-tune and test their evolutionary algorithms. They confirm empirically the benefits of genetic relatedness and group selective pressures for evolving cooperative strategies. Then, they directly apply these properties to the evolution of a complex collective task: the spontaneous formation of wireless communication networks by airborne robots in a mountain rescue scenario. The difficulties of this task should not be underestimated: these robots must build and sustain a communication network adaptively while flying without local position information and measuring their performance only based on their sensed local wireless connectivity. Upon analysis of successful results, the groups of flying robots evolved in simulation yield clear local rules that approximate their observed behavior. The re-implementation of such approximate rules results in reliable global effects, such as the formation of communication chains and network maintenance, thus translating evolved behavior into easier to understand design principles.

More complex behaviors, social or otherwise, in nature and in robots, usually involve (or appear to involve) some kind of higher-level cognitive capabilities (reasoning, decision making, etc). But this is an area where details of natural mechanisms are still very sparse and evolutionary robotics has a potential role to play in both developing artificial cognitive architectures and shedding light on real neural processes. For several decades some of the more prevalent theories of neural mechanisms underlying higher cognitive abilities and complex motor behaviors make use of the notion of compositionality (Arbib 1981; Arbib and Hesse 1986; Tani 2003). The idea is that complex behaviors are built up from simpler reusable primitives that can be combined in a flexible way, with some higher level of control manipulating the behavioral primitives. Although this idea can be quite attractive from a computational perspective, there is not yet clear evidence that this is what actually occurs in real brains. Jun Tani, Michail Maniadakis, and Rainer W. Paine's chapter 10 is a first step toward developing an ER-based methodology for exploring how such mechanisms could be implemented in neural circuits. In the longer term such research could indicate likely classes of biological neural mechanisms, which could then be investigated. They use an evolutionary approach to search for classes of mechanisms that are able to successfully generate compositional goal-directed action generation and rule-switching behaviors in a simulated robot engaged in maze exploration and navigation tasks. This methodology allows them to exploit the fact that fewer assumptions and constraints are needed in the evolutionary approach.

In connection to the challenges of evolving increasingly complex forms of behavior, chapter 11 by Eric D. Vaughan, Ezequiel A. Di Paolo, and Inman Harvey properly illustrates one application of incremental principles as a promising approach. Their task, the evolution of an omni-directional, 3D, minimally actuated, bipedal walker involving several reflex circuits and able to walk robustly over rugged terrain, seems nearly impossible to achieve by attempting to evolve a robot controller from scratch. It seems also quite complex for a human designer. But, the authors show, it is accessible to an interactive scheme involving design and evolution with staged increments in complexity and redesign. The value of their contribution lies both in the actual results as well as in the wider methodological implications of the case study they present.

By breaking down the activity of walking into stages that take into account the roles of the body's passive dynamics, of gravity and of neural reflexes, they are able to evolve circuits that actuate the leg movements of a simulated machine with incremental complexity involving different walking components or "modes." These modes involve, for instance, achieving a certain stance, contracting or swinging legs, maintaining overall support, and so on. Each of these modes is evolved using a relatively simple reflex circuit and these are subsequently integrated. As the complexity of the task is incremented (e.g., moving from flat to sloped terrains or adding degrees of freedom), the earlier stages define the starting point of the new task and much of the circuitry is reused and the lessons learned reapplied. The result is a minimally actuated 3D walking machine with 35 degrees of freedom (in one instantiation) involving several simple, coordinated neural circuits that rely on a strong environmental and bodily feedback.

There are major ongoing challenges—methodological, theoretical, and technological—in all the areas mentioned in this chapter, such as finding the best way to incorporate developmental processes and lifetime plasticity (including learning and automatic self-repair) within the evolutionary framework (Nolfi and Floreano 1999; Lungarella and Berthouze 2002; Lungarella et al. 2003); understanding better what the most useful building blocks are for evolved neurocontrollers (Husbands et al. 2010; Clune, Mouret, and Lipson 2013; Risi and Stanley 2012); and finding efficient ways to scale work on concurrently evolving bodies and brains while continuing to best exploit the complex interplay among body, nervous system, and environment. Jordan Pollack's contribution to this book in chapter 12 nicely lays out his vision of the kinds of approaches that will be needed to meet these challenges. After a lively reminder of the folly of traditional AI's preoccupation with mimicking human-level intelligence, based on the myth of human symbolic conscious reasoning, he introduces the notion of mindless intelligence—the many complex processes to which we could (and sometimes do) ascribe intelligence but that have no mind—from immune systems to social systems and networks, via the many natural and artificial processes of self-organization we are beginning to understand, to the exquisite mechanisms of evolution itself. Coining the term "ectomental," for outside of mind, to refer to such systems and processes, he begins to explore how mechanisms of organization, learning, repair, assembly, reproduction, recognition, and regulation (woven together by evolution) should be regarded as the way forward for AI, probably the only means by which true artificial intelligence could ever be achieved.

This is a vision that Alan Turing, in his guise as an ER pioneer, would recognize and, we like to think, approve of.

References

Arbib, M. A. 1981. Perceptual structures and distributed motor control. In *Handbook of Physiology—The Nervous System II. Motor Control*, ed. V. B. Brooks, 1449–1480. Bethesda, MD: American Physiological Society.

Arbib, M. A., and M. B. Hesse. 1986. *The Construction of Reality*. Cambridge, UK: Cambridge University Press.

Ashby, W. R. 1952. *Design for a Brain*. London: Chapman and Hall.

Barnett, L. 2001. Netcrawling—optimal evolutionary search with neutral networks. In *Proceedings of the 2001 Congress on Evolutionary Computation* (CEC2001), 30–37. Seoul, Korea: IEEE Press.

Beer, R. 1995. A dynamical systems perspective on environment agent interactions. *Artificial Intelligence* 72:173–215.

Beer, R., and J. Gallagher. 1992. Evolving dynamical neural networks for adaptive behaviour. *Adaptive Behavior* 1:94–110.

Beer, R. D. 2003. The dynamics of active categorical perception in an evolved model agent. *Adaptive Behavior* 11:209–243.

Bongard, J. 2011. Morphological change in machines accelerates the evolution of robust behavior. *Proceedings of the National Academy of Sciences of the United States of America* 108 (4): 1234–1239.

Brooks, R. A. 1999. *Cambrian Intelligence: The Early History of the New AI*. Cambridge, MA: MIT Press.

Cangelosi, A., and T. Riga. 2006. An embodied model for sensorimotor grounding and grounding transfer: Experiments with epigenetic robots. *Cognitive Science* 30 (4): 673–689.

Clark, A., J. Moore, J. Wang, X. Tan, and P. McKinley. 2012. Evolutionary design and experimental validation of a flexible caudal fin for robotic fish. *Artificial Life* 13:325–332.

Cliff, D., I. Harvey, and P. Husbands. 1993. Explorations in evolutionary robotics. *Adaptive Behavior* 2:73–110.

Clune, J., J-B. Mouret, and H. Lipson. 2013. The evolutionary origins of modularity. *Proceedings of the Royal Society B* 280: 20122863.

Di Paolo, E. A. 2000. Behavioral coordination, structural congruence and entrainment in a simulation of acoustically coupled agents. *Adaptive Behavior* 8 (1): 25–46.

Floreano, D., P. Husbands, and S. Nolfi. 2008. Evolutionary robotics. In *Springer Handbook of Robotics*, ed. B. Siciliano and O. Khatib, 1423–1451. Berlin: Springer.

Floreano, D., and F. Mondada. 1994. Automatic creation of an autonomous agent: Genetic evolution of a neural-network driven robot. In *From Animals to Animats 3: Proceedings of the Third International Conference on Simulation of Adaptive Behaviour* (SAB94), ed. D. T. Cliff, P. Husbands, J.-A. Meyer, and S. Wilson, 421–430. Cambridge, MA: MIT Press.

Floreano, D., and J. Urzelai. 2000. Evolutionary robots with online self-organization and behavioral fitness. *Neural Networks* 13 (4–5): 431–443.

Gruau, F. 1995. Automatic definition of modular neural networks. *Adaptive Behavior* 3 (2): 151–183.

Gruau, F., and K. Quatramaran. 1997. Cellular encoding for interactive evolutionary robotics. In *Proceedings of the 4th European Conference on Artificial Life*, ed. P. Husbands and I. Harvey, 366–387. Cambridge, MA: MIT Press.

Harvey, I., P. Husbands, and D. Cliff. 1994. Seeing the light: Artificial evolution, real vision. In *From Animals to Animats 3: Proceedings of the Third International Conference on Simulation of Adaptive Behaviour* (SAB94), ed. D. T. Cliff, P. Husbands, J.-A. Meyer, and S. Wilson, 392–401. Cambridge, MA: MIT Press.

Holland, J. H. 1975. *Adaptation in Natural and Artificial Systems*. Ann Arbor: University of Michigan Press.

Husbands, P., and I. Harvey. 1992. Evolution versus design: Controlling autonomous mobile robots. In *Proceedings of the 3rd Annual Conference on Artificial Intelligence, Simulation and Planning in High Autonomy Systems*, 139–146. Los Alimitos, CA: IEEE Computer Society Press.

Husbands, P., I. Harvey, and D. Cliff. 1995. Circle in the round: State space attractors for evolved sighted robots. *Robotics and Autonomous Systems* 15:83–106.

Husbands, P., A. Philippides, P. Vargas, C. Buckley, P. Fine, E. Di Paolo, and M. O'Shea. 2010. Spatial, temporal and modulatory factors affecting GasNet evolvability in a visually guided robotics task. *Complexity* 16 (2): 35–44.

Izquierdo, E., and T. Buhrmann. 2008. Analysis of a dynamical recurrent neural network evolved for two qualitatively different tasks: Walking and chemotaxis. In *Proceedings of the 11th International Conference on Artificial Life*, ed. S. Bullock et al., 257–264. Cambridge, MA: MIT Press.

Izquierdo, E., I. Harvey, and R. D. Beer. 2008. Associative learning on a continuum in evolved dynamical neural networks. *Adaptive Behavior* 16:361–384.

Jakobi, N. 1998. Evolutionary robotics and the radical envelope of noise hypothesis. *Adaptive Behavior* 6:325–368.

Jakobi, N., P. Husbands, and I. Harvey. 1995. Noise and the reality gap: The use of simulations in evolutionary robotics. In *Proceedings of the 3rd European Conference on Artificial Life*, ed. F. Moran et al., 704–720. Berlin: Springer.

Lipson, H., and J. Pollack. 2000. Automatic design and manufacture of robotic lifeforms. *Nature* 406:974–978.

Long, J. 2012. *Darwin's Devices: What Evolving Robots Can Teach Us about the History of Life and the Future of Technology*. New York: Basic Books.

Lungarella, M., and L. Berthouze. 2002. On the interplay between morphological, neural, and environmental dynamics: A robotic case-study. *Adaptive Behavior* 10:223–241.

Lungarella, M., G. Metta, R. Pfeifer, and G. Sandini. 2003. Developmental robotics: A survey. *Connection Science* 15:151–190.

Miglino, O., H. H. Lund, and S. Nolfi. 1995. Evolving mobile robots in simulated and real environments. *Artificial Life* 2 (4): 417–434.

Miglino, O., K. Nafasi, and C. E. Taylor. 1996. Selection for wandering behaviour in a small robot. *Artificial Life* 2 (1): 101–116.

Negrello, M., and F. Pasemann. 2008. Attractor landscapes and active tracking: The neurodynamics of embodied tracking. *Adaptive Behavior* 16:196–216.

Nolfi, S., and D. Floreano. 1999. Learning and evolution. *Autonomous Robots* 7:89–113.

Nolfi, S., and D. Floreano. 2000. *Evolutionary Robotics: The Biology, Intelligence, and Technology of Self-organizing Machines*. Cambridge, MA: MIT Press.

Nolfi, S., D. Floreano, O. Miglino, and F. Mondada. 1994. How to evolve autonomous roots: Different approaches in evolutionary robotics. In *Proceedings of the International Conference on Artificial Life IV*, ed. R. Brooks and P. Maes, 190–197. Cambridge MA: MIT Press.

Parisi, D., and S. Nolfi. 1993. Neural network learning in an ecological and evolutionary context. In *Intelligent Perceptual Systems*, ed. V. Roberto, 20–40. Berlin: Springer.

Philippides, A., P. Husbands, T. Smith, and M. O'Shea. 2005. Flexible couplings: Diffusing neuro-modulators and adaptive robotics. *Artificial Life* 11 (1–2): 139–160.

Port, R., and T. J. van Gelder. 1995. *Mind as Motion: Explorations in the Dynamics of Cognition*. Cambridge, MA: MIT Press.

Quinn, M. 2001. Evolving communication without dedicated communication channels. In *Advances in Artificial Life: Sixth European Conference on Artificial Life* (ECAL 2001), ed. J. Kelemen and P. Sosik, 357–366. Berlin: Springer Verlag.

Quinn, M., L. Smith, G. Mayley, and P. Husbands. 2003. Evolving controllers for a homogeneous system of physical robots: Structured cooperation with minimal sensors. *Philosophical Transactions of the Royal Society of London, Series A: Mathematical, Physical and Engineering Sciences* 361:2321–2344.

Risi, S., and K. Stanley. 2012. An enhanced hypercube-based encoding for evolving the placement, density, and connectivity of neurons. *Artificial Life* 18 (4): 331–363.

Sims, K. 1994. Evolving 3D morphology and behavior by competition. In *Proceedings of the International Conference Artificial Life IV*, ed. R. Brooks and P. Maes, 28–39. Cambridge, MA: MIT Press.

Stanley, K., and R. Miikkulainen. 2002. Evolving neural networks through augmenting topologies. *Evolutionary Computation* 10 (2): 99–127.

Tani, J. 2003. Learning to generate articulated behavior through the bottom-up and the top-down interaction processes. *Neural Networks* 16:11–23.

Tani, J., and S. Nolfi. 1999. Learning to perceive the world as articulated: An approach for hierarchical learning in sensory-motor systems. *Neural Networks* 12:1131–1141.

Turing, A. M. 1950. Computing machinery and intelligence. *Mind* 59:433–460.

Urzelai, J., and D. Floreano. 2001. Evolution of adaptive synapses: Robots with fast adaptive behavior in new environments. *Evolutionary Computation* 9:495–524.

Walter, W. G. 1953. *The Living Brain*. London: Duckworth.

Williams, P. L., R. D. Beer, and M. Gasser. 2008a. An embodied dynamical approach to relational categorization. In *Proceedings of the 30th Annual Conference of the Cognitive Science Society*, ed. B. C. Love, K. McRae, and V. M. Sloutsky, 223–228. Washington, DC: Cognitive Science Society.

Williams, P. L., R. D. Beer, and M. Gasser. 2008b. Evolving referential communication in embodied dynamical agents. In *Artificial Life XI: Proceedings of the Eleventh International Conference on the Simulation and Synthesis of Living Systems*, ed. S. Bullock et al., 702–709. Cambridge, MA: MIT Press.

2 Evolutionary Robotics and Neuroscience

Phil Husbands, Renan C. Moioli, Yoonsik Shim, Andy Philippides, Patricia A. Vargas, and Michael O'Shea

2.1 Introduction

When research in evolutionary robotics (ER) initially took off in the early 1990s, concerns over the brittleness of traditional artificial intelligence (AI) techniques had recently led to a resurgence of interest in artificial neural networks (ANNs). This fact, coupled with the obvious (loose) analogy between robot control systems and biological nervous systems, meant that most ER researchers naturally gravitated toward neuro-control systems (Husbands and Harvey 1992; Beer and Gallagher 1992; Harvey, Husbands, and Cliff 1994; Parisi and Nolfi 1993; Floreano and Mondada 1994). To many of those researchers neural networks also intuitively seemed to be more evolvable than other possible control substrates such as rules or programs—nodes and connections could be gradually changed or added or deleted in a flexible, open-ended way (Harvey, Husbands, and Cliff 1993; Cliff, Harvey, and Husbands 1993; Beer and Gallagher 1992). In addition, from the earliest days, it has been noted that dynamical recurrent varieties of neural networks, many strongly biologically influenced, allow subtle dynamics that can be readily exploited in the generation of robust adaptive behavior (de Garis 1990; Beer and Gallagher 1992; Harvey, Husbands, and Cliff 1993). Hence, from the outset artificial neural networks have been the predominant control system used in ER. Therefore the field has always had at least a tentative link with neuroscience. However, strands of work in which the link is more explicit have existed since the inception of the field and have continued to develop. They are the focus of this chapter.

The two main classes of ER research in which there is a strong tie with neuroscience are: work involving explicitly biologically inspired neural network controllers, often making use of cutting edge neuroscience, and research in which ER is used to develop or explore neural models aimed at answering specific questions in neuroscience. The former class is concerned with biologically inspired technology while the latter uses computational and robotic tools in scientific research. In some cases the boundary between the two classes can become rather blurred: often within a single piece of research both kinds of motivation can be found.

This chapter attempts to sketch a general map of the kind of relationships that exist between ER and neuroscience, using specific examples to illustrate the wide range of interactions between the two fields A number of case studies are used to explore such relationships in more depth.

Section 2.2 lays out the kinds of interactions between ER and neuroscience that will be considered in the rest of the chapter. Following detailed descriptions of a number of case studies, focusing mainly on ongoing research, the chapter concludes with a discussion of open issues and the prospects for such work.

2.2   Relationships between Evolutionary Robotics and Neuroscience

The two main classes of relevant research mentioned previously—the use of neurobiologically inspired neural networks in ER, and the use of ER modeling in neurobiological research—are rather broad, with many variations on each theme. In the following sections the scope of each of these categories will be fleshed out.

With the acceleration in fundamental neuroscientific research that has taken place since about the turn of the century, it has become ever clearer that the central nervous system (CNS) is a far more sophisticated and exotic system than that portrayed in the old-fashioned connectionist electrical network view of neural processing that has dominated the worlds of artificial neural networks and neural modeling (Dayan and Abbott 2001). The emerging picture is one of many interacting adaptive processes operating over different temporal and spatial scales. These new understandings, particularly with reference to behavior generation, provide rich sources of inspiration, at many different levels of abstraction, for the development of artificial nervous systems for robots. A number of examples of such research are described later.

Natural adaptive and intelligent behavior is the result of complex interactions among nervous system, body, and environment. Biological neural systems are embodied and embedded. Because of this there has been a growing interest in using robots, employing on-board neural circuitry, to model aspects of animal behavior. Such a methodology, the argument goes, can give deeper insights into behavior-generating neural mechanisms than disembodied models (Webb 2001; Beer 2003; Seth et al. 2004), as well as fresh perspectives on what it means to be a cognitive agent (Wheeler 2005). Like any modeling enterprise, there are many issues surrounding how to make robotic models, with their inevitable implementation constraints, properly relevant to biological enquiry. For a discussion of such matters, see Webb 2001, Webb 2009, and AB 2009.

Evolutionary robotics has an important role to play in this context as it allows the exploration of whole classes of mechanisms, the automatic creation of working models when there are insufficient details to fully specify a system in advance, and a reduction in "designer bias" (Harvey et al. 2005) through the use of an automatic search process that does not specify in advance what a solution should look like. Hence it has been recognized as a useful tool in investigating biological hypotheses (Husbands et al. 1997; Harvey et al. 2005; Floreano, Husbands, and Nolfi 2008).

2.2.1 Biological Inspiration

Biological brains are often, quite rightly, posited as the most complex systems studied by science. As their mysteries are slowly unraveled, they provide many potential foci of inspiration for developing neuro-influenced controllers for robots. Four of the main such sources of inspiration that have found their way into ER research are:

- Neural architectures
- Intrinsic neural properties
- Signaling modes
- Brain-body-environment dynamics

Each of these is briefly discussed in the remainder of this section. More detailed case studies involving one or more of these elements are covered later in the chapter.

Observations of the ways in which neural circuitry is organized in nature, particularly in relation to motor behaviors in invertebrates, have led to a number of powerful general architectures being adopted in some areas of ER, for instance work on legged locomotion. Such architectures impose constraints on the properties of neurons and the ways in which they can be connected, thus shaping and restricting the evolutionary search space. A pioneering example is the work of Beer and colleagues (Beer, Chiel, and Sterling 1989; Beer et al. 1997), who introduced an architecture for locomotion based on cross-coupled subnetworks, inspired by the cockroach nervous system (figure 2.1). Each leg is controlled by an identical (or near-identical) fully connected subnetwork containing a small number of neurons (typically five or six). These networks are connected to each other by both cross-body and same-side (intersegmental) wiring as illustrated in figure 2.1. Variations on this insect-inspired architecture have proved very successful in the development of locomotion controllers in many types of legged robots, including bipeds, quadrupeds, hexapods and octopods (Jakobi 1998; Kodjabachian and Meyer 1998; Vaughan, Di Paolo, and Harvey 2004; Reil and Husbands 2002). Other ER approaches to locomotion include work that does not make use of central pattern generators (and therefore is less biologically motivated but interesting nonetheless) to generate sidewinding (snake-like) behavior (Tanev, Ray, and Buller 2005; Kuyucu, Tanev, and Shimohara 2012).

Real neurons are highly sophisticated information processing devices, generally orders of magnitude more complex than the crude caricatures employed in artificial neural networks. Their very sophistication means that detailed modeling becomes computationally very expensive, so it is inevitable that abstractions and simplifications should be made in any model or artificial analog. However, by introducing elements

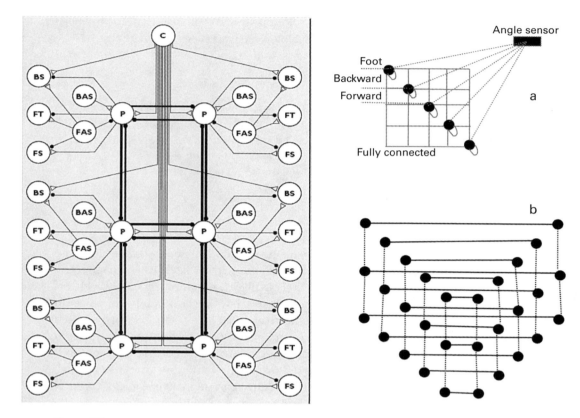

Figure 2.1
Left: schematic diagram of a distributed neural network for the control of locomotion in a hexapod as used by Beer et al. (1989). Excitatory connections are denoted by open triangles, and inhibitory connections are denoted by filled circles. C = command neuron; P = pacemaker neuron; FT = foot motor neuron; FS and BS = forward swing and backward swing motor neurons; FAS and BAS = forward angle sensors and backward angle sensors. Reproduced with permission. Right: generalized architecture using a fully connected dynamical network controller for each leg (a), cross-coupled as shown (b). Solid lines are cross-body connections, and dashed lines are intersegmental connections.

of realistic neural properties, including dynamics, much can be gained. For instance, the particular brand of continuous-time recurrent neural networks (CTRNNs) used to great effect by many ER practitioners (Beer and Gallagher 1992; Beer 2003) are based on simple integrate-and-fire model neurons (Abbott and Kepler 1990), which were originally developed to approximate aspects of neural membrane potential dynamics without the expense of more detailed Hodgkin and Huxley- (1952) style models. In contrast to the uniform nature of most ANNs, individual real neurons and synapses often have their own highly unique properties. In many identified motor circuits, particularly in invertebrates, nerve cells and synapses have widely varying intrinsic properties and behaviors (North and Greenspan 2007). Incorporating such heterogeneity into networks is particularly suited to the ER methodology and has often been successfully employed (Nolfi and Floreano 2000; Floreano, Husbands, and Nolfi 2008). Many other intrinsic properties, such as spiking and oscillatory dynamics (Izhikevich 2007), homeostatic mechanisms (Davis 2006), and numerous forms of plasticity (e.g., Pinaud, Tremere, and De Weerd 2006; Katz 1999) provide a rich vein of inspiration.

Until recently nearly all work in ANNs and computational neuroscience concentrated on the transmission of electrical signals between neurons via axons and dendrites. However, nervous systems are complex electrochemical systems with many non-electrical signaling modes also in play. This fuller picture of neural information processing has inspired work in ER that makes use of analogs of chemical transmission. In particular, the notion of *volume signaling*, whereby neurotransmitters freely diffuse into a relatively large volume around a nerve cell, potentially affecting many other neurons irrespective of whether or not they are electrically connected (Gally et al. 1990; Wood and Garthwaite 1994), has been explored in a body of ER work (Husbands et al. 1998; Philippides, Husbands, et al. 2005; Husbands et al. 2010). This exotic form of neural signaling, which involves modulation of neural or synaptic properties, or both, by the diffusing neurotransmitter, does not sit easily with classical connectionist (point-to-point) pictures of brain mechanisms and is forcing a radical rethink of existing theory (Dawson and Snyder 1994; Philippides, Husbands, and Shea 2000; Philippides, Ott, et al. 2005; Bullock et al. 2005; Katz 1999). Other examples include the use of analogs of neuro-endocrine interactions in ER control architectures (Vargas et al. 2005), neural plasticity modulated by dopamine-inspired mechanisms (Doya 2002), and the introduction of neuron-level homeostatic mechanisms in spiking plastic networks (Di Paolo 2003).

One of the driving forces behind the development of ER and other forms of "New AI" has been the realization that a proper understanding of intelligence must recognize the central role of *embodiment* (Varela, Thompson, and Rosch 1991; Brooks 1991; Clark 1999). The body is not just a passive vessel to be controlled. Rather, adaptive behavior emerges out of a subtle interplay among brain, body, and environment (Pfeifer and Bongard 2007). For instance, studying neural circuitry underlying the generation of rhythmic motor behavior in isolation ignores the considerable advantage that can be obtained from incorporating the physical body and its environment—an approach that can significantly reduce the amount of information needed to develop successful motor patterns. ER has proved to be very useful in exploring and exploiting the interplay among neural, bodily, and environmental dynamics in the development of efficient and robust behaviors (Floreano, Husbands, and Nolfi 2008).

Section 2.3 describes in some detail three examples of current work from the Centre for Computational Neuroscience and Robotics at Sussex University that mainly fall under the biological inspiration category. We say mainly, because although the studies are of biologically inspired neuro-robotic systems, there are elements of biological inquiry in each case.

2.2.2 Biological Modeling

Many types of models are now commonly used in science, including: the classical modeling of a target system at a particular level of abstraction according to some set of assumptions; and models as existence proofs—used to help to refute, or at least cast doubt on, certain claims about necessary conditions for a phenomenon as well as demonstrating new possibilities. Models can also act as substitutes for theories where none exist, a situation common in cognitive science and biology, and, in the case of computational models, as a kind of animated thought experiment aimed at clarifying conceptual issues. ER has been used to develop examples of all of these types of models in relation to various issues in neuroscience; several illustrative cases are discussed later.

A slightly more abstract characterization of the main classes of neurobiological modeling that have employed ER methods, which will be used in the remainder of this chapter, uses the following three categories:

• Model tuning
• Model synthesis
• Development of probing models

Model tuning here refers to adjusting or setting parameter values in an otherwise well-defined model (parameter fitting). There is a growing history of using search methods, including evolutionary techniques, to either fine-tune values or set unknown values in scientific models, including neuroscientific models (e.g., Gerken et al. 2005; Gurkiewicz and Korngreen 2007). When the model is of a whole embodied behavior-generating system, evolutionary robotics can be an ideal tool to use in this context. In that case parameters might describe neural properties or aspects of sensors or bodies, or all of these things.

Model tuning is used when sufficient details of the target system are known to be able to define a parametric model at the desired level of abstraction. In cases where insufficient details are available (e.g., the connectivity of neural circuitry is unclear, or the number of neurons involved is unknown) model synthesis techniques can be employed. ER can be used for this purpose by searching a space of possible models, constrained by available knowledge, to find one that fits the data (e.g., generates appropriate behavior). By attempting to fill in the blanks, a model synthesized in this way presents a set of hypotheses about the target system—for instance, how details of neural architectures and mechanisms underly behavior generation. The hypotheses thus generated can then be subjected to empirical scrutiny in the original biological system. When knowledge of the target system or phenomenon is sparse, such synthesized models generally stand as existence proofs that serve to catalyze further debate and sharpen theories.

A kind of model that is closely related to the existence proof and that does not require any direct representational function, but is used in all branches of science, is the *toy*, or *probing*, model (Frigg and Hartmann 2008). Evolutionary robotics has increasingly been used to develop such models, which operate at a more abstract level than the other types previously listed. They are not intended to represent a specific concrete target system or phenomenon, but to be used as simple vehicles for testing new tools and methods, preparatory to more detailed, empirically based modeling (Hartmann 1995). This is exactly the motivation Randy Beer gives for the strand of research that employs ER techniques to synthesize abstract models of agents engaged in "minimally cognitive" behaviors (Beer 2003). He describes his work on simple autonomous agents involved in categorizing objects within a behavioral context thus: "The intention here is not to propose a serious model of categorical perception, but rather to use this model agent to explore the implications of dynamical explanation for cognitive agents" (210), and exhorts us to "Think of this exercise, then, as a form of mental calisthenics, an intellectual warm-up for the dynamical analyses of a wider range of agents and behaviors" (210). An interesting associate of the toy model is the false model—a model of something known to be wrong—which can have a useful heuristic role in refining and developing "true" models by elaborating their underlying assumptions (Wimsatt 2002).

For an area as difficult and underdeveloped as dynamical analyses and explanations of embodied situated agent behavior, which could have a significant effect on thinking in neuroscience, Beer's justification seems appropriate and pragmatic. The reason toy models are used in physics is the same reason Beer uses them: their relative tractability. ER can be a very powerful tool for building them.

A good example of using ER techniques for model tuning is the work of Ijspeert and colleagues (Ijspeert, Hallam, and Willshaw 1999; Hallam and Ijspeert 2003) on models of the central pattern generator neural circuitry underlying swimming behaviors in the lamprey. The starting point was a model at an intermediate level of abstraction devised by Ekeberg (Ekeberg 1993; Ekeberg, Lansner, and Grillner 1995), which was based on neurophysiological data (Grillner, Wallen, and Brodin 1991). This model used networks of simple leaky integrator nodes to control a simulated lamprey body. The nodes in the networks can be thought of as populations of neurons and the connections between them as general pathways in the lamprey's spinal cord. This model had a well-defined set of parameters that described individual node and connection properties. Ekeberg hand-designed a set of specific parameter values that were able to reproduce some of the real lamprey's behaviors to a good level of accuracy (Ekeberg, Lansner, and Grillner 1995). Ijspeert and colleagues used a genetic algorithm to search the parameter space of this model and were able to find several combinations different from Ekeberg's hand-designed set that reproduced the biological data better. Since the ER-generated solutions were all essentially variations of Ekeberg's model, the work also showed that the original model was fairly robust to differences in parameter values (Hallam and Ijspeert 2003). Ijspeert also used ER as a model synthesis technique by relaxing the constraints imposed by Ekeberg's parametric model and searching the resulting space of possible models. They found a number of alternative models that were also able to reproduce the biological data. In a related vein, von Twickel, Büschges, and Pasemann (2011) have used a mixture of hand coding and ER methodology to develop single-leg model controllers based on empirical observations of stick insect neurobiology. They were able to reproduce a range of behaviors that matched the biological data with their neural models.

Another recent example of using ER for model synthesis is the work of Izquierdo and Lockery (2010). They used ER methods to develop a model of the neural mechanisms underlying klinotaxis, a common form of chemotaxis, in *C. elegans*. Previous neural models of chemotactic behavior in nematodes focused on the other strategy they commonly use: klinokinesis, in which the direction of movement is governed by a biased random walk (Ferree and Lockery 1999). Klinotaxis involves movements in which the direction of locomotion in a chemical gradient closely follows the line of steepest ascent. The differences between the two forms of behavior imply a distinctive neural network controlling klinotaxis. This network has not yet been identified and no hypothetical model existed before Izquierdo and Lockery's work. They used an evolutionary algorithm to generate neural networks that exhibited klinotaxis in a simple idealized physical model of *C. elegans*. Sensory inputs and motor outputs of the model networks were constrained to match empirical data as were other aspects of a hypothesized network architecture. Motor neurons were modeled as simple leaky integrators as used in CTRNNs (Beer 1995). The parameters of the resulting network were evolved to discover working instances of the network that could then stand as hypotheses about the mechanisms at play in *C. elegans*. They discovered that a minimalistic neural network, comprised of an ONOFF pair of chemosensory neurons and a pair of neck-muscle motor neurons, is sufficient to generate realistic klinotaxis behavior. Importantly, emergent properties of model networks reproduced other experimental observations that they were not designed to fit, suggesting that the model may be operating according to principles similar to those of the biological network. A large number of successful networks were analyzed and this revealed a novel neural mechanism to allow asymmetric turning behavior (a kind of mutual inhibition between motor neurons is achieved simply by shifting the sigmoidal input–output function of the motor neurons relative to the dynamic range of the oscillatory input driving the nematode's head sweeps). The authors stress that this mechanism provides a testable hypothesis that is likely to accelerate the discovery and analysis of the biological circuitry for chemotaxis in *C. elegans*.

Seth (2005) gives an interesting example of ER used in a probing model context. The aim of the work was to develop methods based on Granger causality (Granger 1969) to enable analysis of causal interactions occurring within behavior-generating neural mechanisms. This requires detailed simultaneous temporal data from several sites in some relevant neural circuitry, in practice not at all easy to gather from an intact biological neural system. Rather than try and develop the method (termed "causal connectivity analysis") using data somehow collected from a biological context, Seth very sensibly opted to experiment with the ideas in an abstract simulation. This is a classic case of the kind of situation in which probing models can be very useful, as discussed earlier. He used a genetic algorithm to develop model neural networks optimized for controlling target fixation in a simulated head–eye system, in which the structure of the environment could be experimentally varied. Causal connectivity analysis of a number of networks evolved in various contexts within this framework gave novel insights into neural mechanisms underlying sensorimotor coordination. Seth demonstrated that networks underlying relatively rich adaptive behavior showed a higher density of causal interactions, as well as a stronger causal flow from sensory inputs to motor outputs, than networks generating relatively simple behaviors. In addition he showed that this style of analysis can predict the functional consequences of network lesions. The methods developed using this probing model were powerful enough to suggest that causal connectivity analysis, and similar techniques, could have useful applications in the analysis of real neural dynamics, a line of inquiry that has been followed by Seth and colleagues with some success (e.g., Seth, Barrett, and Barnett 2011, Barnett and Seth 2011).

Suzuki, Floreano, and DiPaolo's (2005) work on whether or not proprioceptive motor information resulting from the generation of actions is necessary for the development of normal, visually guided behavior is another example of ER being used to explore an explicitly neuroscientific question. In an experiment inspired by Held and Hein's (1963) work on cats, two initially identical evolved robots were compared. One was left free to move in a square environment while the other was forced to move along trajectories imposed externally, but was free to control its camera position. The visual receptive fields and behaviors of the passive robot significantly differed from those of the active robot. Further analysis revealed that passive robots became oversensitive to features that were not functional to their normal behavior and which interfered with other dominant features in the visual field. This lead to a hypothesis that some pathological behaviors seen in animals might have roots in similar developmental deficiencies.

Tani, Maniadakis, and Paine (chapter 10, this volume) present another example of ER used to develop abstract probing models as a first step to shed light on possible neural mechanisms involved in higher-level cognitive processes, such as compositional goal-directed action generation and rule switching.

2.3   Neuroscience-Inspired ER Case Studies

This section illustrates in more detail how ideas from empirical and theoretical neuroscience can provide powerful inspiration for work in ER by focusing on three examples of current work from our lab at Sussex University.

2.3.1   Volume Signaling: GasNets

A good example of ER research drawing strongly on inspiration from neuroscience concerns the class of artificial neural networks developed to explore an analog of volume signaling—so-called GasNets (Husbands et al. 1998). These take particular inspiration from nitric oxide (NO) signaling (Gally et al. 1990). They comprise a fairly standard artificial neural network augmented by a chemical signaling system based on a diffusing *virtual* gas that can modulate the response of other neurons. Because there was (and still is) insufficient knowledge of the biological systems to completely define artificial systems working on similar principles, Husbands and colleagues developed the networks to be used within an evolutionary robotics context (Husbands et al. 1998). Thus researchers at Sussex University (Philippides, Husbands, et al. 2005; Husbands et al. 2010) have explored a number of GasNet variants, inspired by different aspects of real nervous systems, as artificial nervous systems for mobile autonomous robots. These variants are being investigated as potentially useful engineering tools, including as modules in complex robot control systems (Vargas et al. 2009), while a related strand of more detailed modeling work is aimed at gaining helpful insights into biological systems (Philippides, Husbands, and O'Shea 2000; Philippides et al. 2003; Philippides, Ott, et al. 2005).

By analogy with biological neuronal networks, GasNets incorporate two distinct signaling mechanisms, one "electrical" and one "chemical." The underlying "electrical" network is a discrete time step, recurrent neural network with a variable number of nodes. These nodes are connected by either excitatory or inhibitory links with the output, O_i^t, of node i at time step t determined by the following equation:

$$O_i^t = \tanh\left[k_i^t \left(\sum_{j \in \Gamma_i} w_{ji} O_j^{t-1} + I_i^t \right) + b_i \right] \tag{2.1}$$

where Γ_i is the set of nodes with connections to node i and $w_{ji} = \pm1$ is a connection weight. I_i^t is the external (sensory) input to node i at time t, and b_i is a genetically set bias. Each node has a genetically set default transfer function gain parameter, k_i^0, which can be altered at each time step according to the concentration of diffusing "gas" at node i to give k_i^t (as described later).

In addition to this underlying network in which positive and negative "signals" flow between units, an abstract process loosely analogous to the diffusion of gaseous modulators is at play. Some units can emit virtual "gases," which diffuse and are capable of modulating the behavior of other units by changing their transfer functions. The networks occupy a 2D space; the diffusion processes mean that the relative positioning of nodes is crucial to the functioning of the network. Spatially, the gas concentration varies as an inverse exponential of the distance from the emitting node with spread governed by a parameter, r, genetically set for each node, which governs the radius of influence of the virtual gas from the node as described by the equations that follow and as illustrated in figure 2.2. The maximum concentration at the emitting node is 1.0 and the concentration builds up and decays linearly as dictated by the time course function, $T(t)$, defined as follows.

$$C(d,t) = \begin{cases} e^{-2d/r} \times T(t) & d < r \\ 0 & \text{else} \end{cases} \qquad (2.2)$$

$$T(t) = \begin{cases} 0 & t = 0 \\ min\left(1, \left(T(t-1) + \dfrac{1}{s}\right)\right) & \text{emitting} \\ max\left(0, \left(T(t-1) - \dfrac{1}{s}\right)\right) & \text{not emitting} \end{cases} \qquad (2.3)$$

where $C(d,t)$ is the concentration at a distance d from the emitting node at time t and s (controlling the slope of the function T) is genetically determined for each node. The range of s is such that the gas diffusion timescale can vary from 0.5 to 0.09 of the timescale of "electrical" transmission (i.e., a little slower to much slower). The total concentration at a node is then determined by summing the contributions from all other emitting nodes (nodes are not affected by their own emitted gases to avoid runaway positive feedback). The diffusion process is modeled in this simple way to provide extreme computational efficiency, allowing arbitrarily large networks to be run very fast—a very useful property in the context of evolutionary search.

For mathematical convenience, in the original basic GasNet there are two "gases," one whose modulatory effect is to increase the transfer function gain parameter (k_i^t) and one whose effect is to decrease it. It is genetically determined whether or not any given node will emit one of these two gases (gas 1 and gas 2), and under what

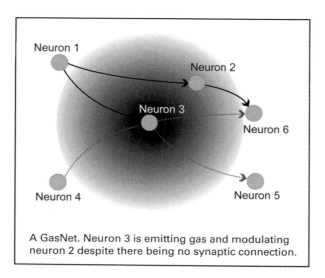

Neuron 1

Neuron 2

Neuron 3

Neuron 6

Neuron 4

Neuron 5

A GasNet. Neuron 3 is emitting gas and modulating neuron 2 despite there being no synaptic connection.

Figure 2.2

A basic GasNet showing excitatory (solid) and inhibitory (dashed) "electrical" connections and a diffusing virtual gas creating a "chemical" gradient.

circumstances emission will occur (either when the "electrical" activation of the node exceeds a threshold, or the concentration of a genetically determined gas in the vicinity of the node exceeds a threshold; note these emission processes provide a coupling between the electrical and chemical mechanisms). The concentration-dependent modulation is described by the following equation, with transfer function parameters updated on every time step as the network runs:

$$k_i^t = k_i^0 + \alpha C_1^t - \beta C_2^t \tag{2.4}$$

where k_i^0 is the genetically set default value for k_i, C_1^t and C_2^t are the concentrations of gas 1 and gas 2 respectively at node i on time step t, and α and β are constants such that $k_i^t \in [-4, 4]$. Thus the gas does not alter the electrical activity in the network directly but rather acts by continuously changing the mapping between input and output for individual nodes, either directly or by stimulating the production of further virtual gas.

The general form of diffusion is based on the properties of a (real) single source neuron as modeled in detail in Philippides, Husbands, and O'Shea (2000) and Philippides et al. (2003). The modulation chosen is motivated by what is known of NO modulatory effects at synapses (Baranano, Ferris, and Snyder 2001). For further details, see Husbands et al. 1998, Philippides, Ott, et al. 2005, and Husbands et al. 2010.

When they were first introduced, GasNets were demonstrated to be significantly more evolvable than a variety of standard ANNs on some noisy, visually guided evolutionary robotics tasks (Husbands 1998; Husbands et al. 1998). Typically the increase in evolvability, in terms of number of fitness evaluations to a reliable good solution, was an order of magnitude or more. The solutions found were often very lean with few nodes and connections, typically far fewer than were needed for other forms of ANN (Husbands et al. 1998). But the action of the modulatory gases imbued such networks with intricate dynamics: they could not be described as simple. Oscillatory subnetworks based on interacting "electrical" and "gas" feedback mechanisms acting on different timescales were found to be very easy to evolve and cropped up in many forms, from CPG circuits for locomotion (McHale and Husbands 2004) to noise filters and timing mechanisms for visual processing (Husbands et al. 1998, Smith et al. 2002). GasNets appeared to be particularly suited to noisy sensorimotor behaviors, which could not be solved by simple reactive feedforward systems, and to rhythmical behaviors.

Two recent extensions of the basic GasNet, the receptor and the plexus models, incorporated further influence from neuroscience (Philippides, Husbands, et al. 2005). In the receptor model, modulation of a node is now a function of gas concentration and the quantity and type of receptors (if any) at the node. This allows a range of site-specific modulations within the same network. In the plexus model, inspired by a type of NO signaling seen in the mammalian cerebral cortex (Philippides, Ott, et al. 2005), the emitted gas "cloud," which now has a flat concentration, is no longer centered on the node controlling it but at a distance from it. Both these extended forms proved to be significantly more evolvable again than the basic GasNet. Other varieties include nonspatial GasNets where the diffusion process is replaced by explicit gas connections with complex dynamics (Vargas et al. 2009) and versions with other forms of modulation and diffusion (Husbands et al. 2010). In order to gain insight into the enhanced evolvability of GasNets, detailed comparative studies of these variants with each other, and with other forms of ANN, were performed using the robot task illustrated in figure 2.3 (Philippides, Husbands, et al. 2005; Husbands et al. 2010).

The question naturally arises as to why the GasNet and variants are more evolvable. Intriguingly, in a comprehensive study Smith, Husbands, and O'Shea (2003) found no explanation for increased GasNet evolvability in terms of fitness landscape properties (neutrality, epistasis, etc.), apart from at high fitness values. Smith and colleagues argued that the key to understanding the improvement of the GasNet was to analyze its behavior at a higher level of abstraction. In particular, they showed how the temporal dynamics of the GasNet seemed to make it relatively easy to tune the networks to the timescales needed in the task (Smith et al. 2002). Similar high-level analyses of the spatial structure of successful GasNets and variants led to the hypothesis that it was the level of coupling between the electrical and gas signaling systems that was

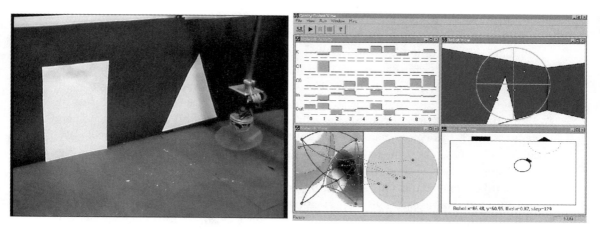

Figure 2.3
Left: the gantry robot. A CCD camera head moves at the end of a gantry arm allowing full 3D movement. In the study referred to in the text, 2D movement was used, equivalent to a wheeled robot with a fixed forward-pointing camera. A validated simulation was used: controllers developed in the simulation work at least as well on the real robot. Right: the simulated arena and robot. The bottom-right view shows the robot position in the arena with the triangle and rectangle. Fitness is evaluated on how closely the robot approaches the triangle. The top-right view shows what the robot "sees," along with the pixel positions selected by evolution for visual input. The bottom-left view shows how the genetically set pixels are connected into the control network whose gas levels are illustrated. The top-left view shows current activity of nodes in the GasNet.

key. In particular that successful evolution came through the systems being flexibly coupled: neither independent of each other nor too tightly bound, allowing one system to be "tuned" against the other without causing catastrophic destructive interference (Philippides, Husbands, et al. 2005). Throughout, however, it was clear that these factors did not act in isolation and that it is the modulatory effect of the gas that lends the networks their adaptivity. This leads to three linked hypotheses on why the GasNets evolve faster:

• The action of gas over multiple different timescales from the electrical activity introduces rich dynamics that can be exploited.
• The spatial embedding of the networks serves to (flexibly) couple two interacting signaling systems.
• The particular modulatory effects are key to evolvability.

These hypotheses were examined in an extended empirical study, discussed in the section that follows, which compared variants of the basic GasNet formed by imposing various constraints on spatial, temporal, and modulatory properties.

Table 2.1

The GasNet variants used in a comparative study

#	Name	Description
1	gnet	Basic gasnet as described in a previous section (2.3.1).
2	nchem	Basic gasnet but with all chemicals inactive.
3	gnetN	Basic gasnet with no diffusion dynamics, i.e., $T(t) = 1$, for all t (see equation 2.3).
4	gnetNw	The same as gnetN but with $T(t) = w$ where $w \in \{0, 1, 2\}$ is a "gas weight" genetically set for each node.
5	flatR	The same as gnet except the gas concentration within the genetically set radius for each emitter is flat with no gradient (the term $e^{-2d/r}$ in equation 2.2 is replaced by e^{-1}).
6	flatRN	The same as flatR except without diffusion dynamics, i.e., $T(t) = 1$, for all t (see equation 3).
7	flatE	The same as flatR except the influence of the gas is not confined to the genetically set radius of influence for a node but now extends everywhere.
8	flatEN	The same as flatE but without diffusion dynamics.
9	AddMod	The most radical variant where the multiplicative modulation of the basic GasNet is replaced by an additive modulation as described by equation 2.5 (i.e., the gas no longer modulates the transfer function gain parameter but instead modulates an additional additive bias term).

Comparative Study

Nine GasNet variants were compared in order to probe the hypotheses about GasNet evolvability outlined earlier. These variants implement a range of constraints affecting spatial, temporal, and modulatory factors, and are described in table 2.1.

$$O_i^t = \left[k_i^0 \left(\sum_{j \in \Gamma_i} w_{ji} O_j^{t-1} + I_i^t \right) + b_i + \gamma_i \left(C_1^t - C_2^t \right) \right] \tag{2.5}$$

The task used in the studies is illustrated in figure 2.3. Starting from an arbitrary position and orientation in a black-walled arena, a robot equipped with a forward-facing camera must navigate under extremely variable lighting conditions to one shape (a white triangle) while ignoring the second shape (a white rectangle). The robot must successfully complete the task over a series of trials in which the relative position and size of the shapes vary. Both the robot control network and the robot sensor input morphology, that is, the number and positions of the camera pixels used as input and how they were connected into the network, were under evolutionary control as shown in figure 2.3. The network architecture (including number of nodes) and all properties of the nodes and connections and gas diffusion parameters were set by an evolutionary search algorithm. Because of the noise and variation, and limited sensory capabilities (only very few pixels are used), this task is challenging, requiring robust, general solutions. The gantry robot shown in the figure was used. Evolution took place in a special validated simulation of the robot and its environment.

In all cases networks were encoded on a variable-sized genotype coding for a variable number of nodes. A genotype consisted of an array of integer variables, each lying in the range [0,100]. For continuous variables, the phenotype value is obtained by normalizing the genotype value to lie in the range [0.0,1.0] and multiplying by the relevant variable range. For nominal values, such as whether or not the node has a visual input, the phenotype value = genotype value MOD N_{nom}, where N_{nom} is the number of possible nominal values, and MOD is the binary modular division operator. Each node in the network had between 19 and 21 variables associated with it, depending on which network variant it described. These define the node's position on a 2D plane; how the node connects to other nodes on the plane with either excitatory (weight +1) or inhibitory (weight –1) connections; whether or not the node has visual input, and if it does the coordinates of the camera pixel it takes input from, along with a threshold below which input is ignored; whether or not the node has a recurrent connection; whether and under what circumstances the node can emit a gas and if so which gas it emits; and a series of variables describing the gas emission dynamics (maximum range, rate of emission and decay, etc.). All variables were under evolutionary control. Four of the nodes are assigned as motor nodes (forward and backward nodes for the left and right motor, with motor speeds proportional to the output of the relevant forward node minus the output of the relevant backward node). See Husbands et al. 1998, 2010 for full details.

Sixteen evaluations were carried out on an individual network, with scores f_i calculated on the fraction of the initial robot-triangle distance that the robot moves toward the triangle by the end of the evaluation; a maximum score of 1.0 is obtained by getting within 10.0cm of the triangle at any time during the evaluation. The controller only receives visual input; reliably getting to the triangle over a series of trials with different starting conditions, different relative positions of the triangle and rectangle, and under very noisy lighting, can only be achieved by visual identification of the triangle. The evaluated scores are ranked, and the fitness F is the weighted sum of the N=16 scores, with weight proportional to the *inverse* ranking i (ranking is from 1 to N, with N as the *lowest* score):

$$F = \frac{\sum_{i=1}^{N} i f_i}{\sum_{i=1}^{N} i} = \frac{2}{N(N+1)} \sum_{i=1}^{N} i f_i \qquad (2.6)$$

Note the higher weighting on the poorer scores provides pressure to do well on *all* evaluations; a solution scoring 50 percent on every evaluation has fitness nearly four times that of one scoring 100 percent on half of the evaluations and zero on the other half.

A geographically distributed, asynchronous updating evolutionary algorithm was used (Collins and Jefferson 1991; Husbands 1992; Husbands et al. 1998), with a population size of 100 arranged on a 10×10 grid. Parents were chosen through rank-based roulette-wheel selection on the mating pool consisting of the eight nearest neighbors to a randomly chosen grid-point. A mutated copy of the parent was placed back in the mating pool using inverse rank-based roulette-wheel selection. Three mutation operators were applied to solutions during evolution. Each *integer* in the genotype string had a 10 percent probability of mutation in a Gaussian distribution around its current value (for certain genes, 20 percent of its mutation will be random jumps within the full possible range). There was also an addition operator, with a 4 percent chance per *genotype* of adding one neuron to the network by inserting a block of random values describing each of the new node's properties, and a deletion operator, also with a 4 percent chance per *genotype* of deleting one randomly chosen neuron from the network. An evolutionary run is terminated when a perfect score has been achieved in 10 successive generations, or after 10,000 generations if the former criteria are not met.

Results of the comparative study are summarized in figure 2.4. A quick glance suggests that the basic GasNet (group 1) is the most consistently evolvable with group 9 (AddMod) clearly the worst (no runs were successful). Group 2 (nchem), in which gas effects are turned off, performs poorly on most runs, although, like most other variants, some runs produce good solutions relatively quickly. Most other network types without diffusion dynamics, thereby robbed of rich temporal properties, including multiple timescales, perform relatively poorly (groups 3, 4, 6). However, the relatively good performance of group 8, without dynamics, especially compared to group 7, which has dynamics, suggests that the story is not quite as simple as it might at first appear. Since it is not possible to assume the data distributions are normal, nonparametric statistical procedures were used to test for significant differences between the network types. A Kruskal-Wallis test performed on the whole data set (all nine groups) revealed highly significant differences between the distributions ($p < 10-14$). Pair-wise Wilcoxon Rank-Sum tests, adjusted for multiple comparisons using the Dunn-Sidak procedure for controlling type-1 statistical error, were used to further probe the differences between the distributions. These tests showed that all network types, except group 6 (flatRND), were significantly more evolvable (in terms of generations to consistent success) than group 9 (AddMod). Since the Dunn-Sidak procedure is necessarily conservative and becomes more so as the number of groups increases, pairwise comparisons were recalculated for all network types except AddMod (i.e., groups 1–8). The results of these comparisons are shown in table 2.2.

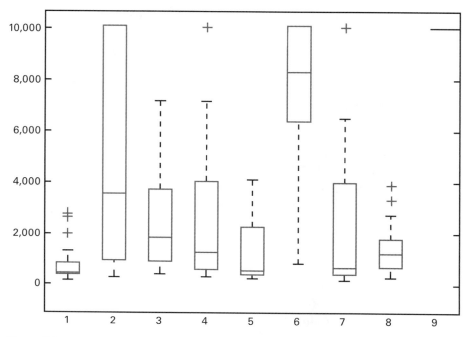

Figure 2.4

Boxplot summarizing results of the comparative study. The x-axis refers to the network type numbers as shown in table 2.2. The y-axis shows generations to success as defined by the stopping criteria explained in the text. The horizontal line within each box is the median, the top and bottom of the box show the 75th and 25th percentiles respectively, the whiskers extend to extreme points of the data not considered outliers (as defined by Rosner's test), with outliers plotted individually. Forty runs of each network type are included.

The results reveal the importance of the dynamics conferred by the diffusing gases. The basic GasNet (group 1) is significantly more evolvable than the variant with the gas turned off (group 2) as well as the variants with the gas operating but without dynamics (groups 3 and 4). It is also significantly more evolvable than the variant with the gas operating but with neither a concentration gradient nor dynamics (group 6).

However, there is one group without gas dynamic than which the basic GasNet is not significantly more evolvable: group 8 (flatEN), of which more later. The version of the GasNet with diffusion dynamics but without a concentration gradient (group 5, flatR) performs fairly well with a low minimum and median, but the fairly high spread of results means that it is not as reliably evolvable as the basic GasNet. There is a similar story for group 7 (flatE) but its reliability is even worse; it should be noted

Table 2.2
Summary of tests for differences between evolvability (generations to consistent success). Distributions for network types 1–8 were tested against each other using pair-wise Wilcoxon Rank-Sum tests adjusted for multiple comparisons using the Dunn-Sidak procedure. Cell entries state whether or not there is a significant difference between the two distributions in question ($p < 0.05$).

Sig diff?	1	2	3	4	5	6	7	8
1	n	Y	Y	n	n	Y	n	n
2	Y	n	n	n	n	n	n	n
3	Y	n	n	n	n	n	n	n
4	n	n	n	n	n	Y	n	n
5	n	n	n	n	n	Y	n	n
6	Y	n	n	Y	Y	n	Y	Y
7	n	n	n	n	n	Y	n	n
8	n	n	n	n	n	Y	n	n

that this restricted form of GasNet has similarities with various network models of neuromodulation that use global modulator signals (Doya 2002).

Although these results suggest there is more to the GasNet's evolvability than the multiple timescales provided by the gas diffusion dynamics, they do add a certain amount of weight to previous suggestions (Philippides, Husbands, et al. 2005) that their easily tuneable dynamics is an important part of their success (as well as to more general claims about the importance of dynamics in the generation of behavior).

Even more obvious is the role of the type of modulation used—additive modulation proved to be useless (group 9). The multiplicative modulation employed in all other variants is able to assert a much more drastic influence on a node, being able to radically change the transfer function by altering the gain k_i^t (equation 2.1)—for instance, flipping the slope from positive to negative or making it flat. These kinds of radical changes were dynamically employed in most successful GasNets and were at the heart of mechanisms, such as oscillators, used to produce stable reliable behavior in the face of significant noise (Husbands et al. 1998; Smith et al. 2002). Additive modulation, which acts at the same level as a node input or bias, could not produce strong enough effects to generate stable behavior. When GasNets were first introduced (Husbands 1998) an alternative node transfer function was successfully used along with an exponential modulation (changing exponents in a polynomial transfer function) that allowed potentially large alterations to the transfer function, which seems to be necessary for effective evolution. These kinds of (multiplicative or exponential) modulations may well confer evolutionary advantages by allowing network nodes to be sensitive to different ranges of input (internal and sensory) in different contexts. For instance, in one (behavioral) context an input node may need to be sensitive to a range of low sensor values while in another it is required to be sensitive to a range of high values. Changing a node's gain through multiplicative modulation allows its sensitivity to be adjusted in an appropriate way.

The spatial embedding of the networks also appears to play a role in producing the most effective coupling between the two distinct signaling processes ("electrical" and "chemical"). By exploiting a loose, flexible coupling between the two processes, it is possible to significantly reduce destructive interference between them, allowing one to be "tuned" against the other while searching for good solutions. It has been suggested that similar forces may be at play in spiking networks, where subthreshold and spiking dynamics interact with each other, which have been evolved to drive vision-based robot behaviors (Floreano and Mattiussi 2001; Floreano, Husbands, and Nolfi 2008). In the most successful varieties of GasNet, dynamics, modulation, and spatial embedding act in concert to produce highly evolvable degenerate (Tononi, Sporns, and Edelman 1999) networks.

2.3.2 Coupled Oscillator Networks and Minimal Cognition

From shortly after the birth of modern neuroscience at the turn of the last century, researchers have looked at neuronal dynamics from an oscillatory perspective (Berger 1929). The consensus nowadays is that cognitive processes have a close nontrivial relationship to neuronal rhythms and oscillations (Buzsaki 2006). In recent years various researchers have stressed the importance of considering temporal relations among groups of neurons, modulated by external influences or sustained by internal mechanisms or both (Engel, Fries, and Singer 2001; Singer 1999; Konig, Engel, and Singer 1996). According to Varela et al. (2001), it is essential to investigate the temporal dynamics of neural networks in order to understand the emergence and integration of neuronal assemblies by means of synchronization. These dynamic assemblies, which are related to large-scale neuronal integration, can influence every cognitive act an agent might eventually perform. In studying these temporal dynamics, Varela and collaborators opted to focus on the phase relationships of brain signals, mainly because these contain a great deal of information on the temporal structure of neural signals, particularly those relating to the underlying mechanism for brain integration. Other authors have emphasized the relationship between phase information and memory formation and retrieval (Li and Hopfield 1989; Izhikevich 1999).

In robotics, although there has been much work on neurally inspired, coupled oscillator-based control of complex rhythmic motor behaviors, particularly locomotion (e.g., Ijspeert et al. 2005), to date there has been very little research on the wider issues of neuronal synchronization and phase information in the generation of embodied cognitive behaviors. The study described next is the first attempt to investigate the neural dynamics of a simulated robotic agent engaged in minimally cognitive tasks while employing an evolved instance of the Kuramoto model of coupled oscillators (Kuramoto 1984) as its nervous system. These tasks are simple enough to allow detailed analysis and yet are complex enough to motivate some kind of cognitive interest. The work has dual aims: first, to shed new light on the possible role of neuronal synchronization and phase information in the generation of sensorimotor cognitive behaviors—for instance to investigate whether different degrees of synchronization are appropriate in different circumstances and what role nonsynchronized, transient dynamics might play—and second, to begin investigating the efficacy of such systems as practical robotic controllers.

The first task is an active categorical perception problem (Beer 2003; Dale and Husbands 2010) in which the robot has to discriminate between moving circles and squares, as first introduced by Beer (2003). In the second task, the robotic agent has to approach moving circles with both normal and inverted vision, adapting to both conditions. Even though these tasks don't strictly require a network of coupled oscillators to be solved, they have been chosen because they are useful benchmarks in the evolutionary robotics and adaptive behavior communities (Di Paolo 2000; Izquierdo 2006) and act as a suitable focus for the possible roles of synchronization in a deliberately nonrhythmic behavior with some relevance to cognition.

The rationale behind the choice of the Kuramoto model is that it describes the phase evolution of a set of connected oscillators and with some adjustments can be associated with groups of neurons firing at a periodic rate (Cumin and Unsworth 2007). Therefore, instead of focusing on single neuron activations, the model resembles the behavior of groups of neurons. By using the phase dynamics as the central feature of the model, the emphasis is on short-term temporal activity, which has been previously shown to be successful in pattern recognition tasks (Tononi Sporns, and Edelman 1992). Moreover, the model allows for easy inspection of the phase and frequency of each of the elements, which makes it especially suitable for studying synchronization of groups of oscillators (Acebron et al. 2005), a key factor when analyzing communication and information processing in neuronal assemblies (Von der Malsburg 1981; Friston 2000). Izhikevich (1999) shows that depending on changes in phase relationships caused by external/internal stimulus, neurons can reorganize and synchronize themselves with different neurons, thus changing their response without the need to change synaptic weights. This points toward new kinds of behavior-generating mechanisms that are explored in the study that follows, based on the Kuramoto model.

The Kuramoto Model

The Kuramoto model consists of a lattice of oscillators coupled according to equation 2.7.

$$\frac{d\theta_i}{dt} = \omega_i + \frac{k}{n}\sum_{j=1}^{n}\sin(\theta_j - \theta_i) \qquad\qquad (2.7)$$

where θ_i is the phase of the ith oscillator, ω_i is the natural frequency of the ith oscillator, k is the coupling factor between nodes, and n is the total number of oscillators. If the frequency of any two given nodes i and j are equal, i.e., $d\theta_i - d\theta_j = 0$ or $\theta_i - \theta_j = constant$, the model is said to be globally synchronized.

It is possible to define a synchronization index, which calculates how synchronized the set of oscillators are (Kuramoto 1984). A commonly used measure is defined in equation 2.8, where r is the synchronization index (a value of 1 indicating high synchronization, 0 indicating incoherent oscillatory behavior) and ψ is the mean phase of the system.

$$re^{i\psi} = \frac{1}{n}\sum_{j=1}^{n}e^{i\theta_j} \qquad\qquad (2.8)$$

The Kuramoto model has a set of properties that makes it suitable for the study of different types of synchronization problems. The work described here focuses on a particular property known as partial synchronism. This phenomenon is exhibited when, in a globally synchronized network, changing the frequency of one node results in some of the nodes become synchronized while other nodes are not (Monteiro et al. 2003).

Moreover, the oscillatory behavior of one node can be influenced by another node in the network not necessarily connected to it. The importance of this property in mimicking brain-related dynamics relies on the fact that different neuronal blocks could synchronize and influence other blocks (e.g., different cortical areas could flexibly establish communication channels depending on their temporal activity). This is in agreement with some recent findings in neuroscience (Buzsaki 2006), reinforcing the feasibility of applying the Kuramoto model to study cognitive processes.

Experimental Framework

The Kuramoto model was modified so that it could be used to control a simulated robotic agent. The core of the robot controller is a set of oscillators, connected in two possible ways: to their two adjacent neighbors, producing a ring structure (see figure 2.5), or fully connected. In his original work, Kuramoto suggested a fully connected setup, but other structures, including the ring-shaped one, have been studied and proven to have a direct influence over the synchronization properties of the model (Wiley, Strogatz, and Girvan 2006; Cumin and Unsworth 2007; El-Nashar and Cerdeira 2009).

The frequency of each node is calculated as the sum of its natural frequency, ω_i, and the value of the sensory input related to that node is scaled by a factor z_i. At each

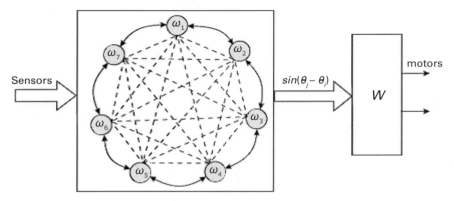

Figure 2.5
Framework for application in evolutionary robotics. The oscillatory network is represented by a set of nodes connected by a thick line, in the case of the ring topology, or by dashed lines, in the case of the fully connected topology.

iteration the phase differences from a given node in relation to the set of all other nodes it is connected to (C_i) are calculated according to equation 2.9. Based on the approach suggested by Schmidhuber et al. (2007), the output of the network is given by the sine of the phase differences linearly combined by an output weight matrix W (figure 2.5). The sine function smooths out phase difference instabilities caused by phases resetting when they exceed 2π. Therefore there are n inputs to n corresponding nodes in the network, with $C_{n,2}$ resulting phase differences and o outputs created via a $C_{n,2} \times o$ matrix W.

$$\frac{d\theta_i}{dt} = (w_i + z_i I(t)) + k \sum_{j \in C_i} \sin(\theta_j - \theta_i) \qquad (2.9)$$

In this way, the overall behavior of the network will be dictated by the phase dynamics and the environmental input to the robotic agent. It is important to stress that nodes that are not directly connected can still influence each other, depending on their frequencies.

The first experiment involved an active categorical perception task performed by a simulated circular robotic agent able to move horizontally at the bottom of a 250 × 200 rectangular environment (figure 2.6). The robotic agent has seven ray sensors, symmetrically displaced in relation to the central ray in intervals of $\pm\pi/12$ radians, and two motors. An intersection between a sensory ray and an object gives a sensor reading between 0 and 10, 0 when the ray length is greater than 200 units and 10 when the ray length is 0. In all experiments, sensors are saturated (they are clamped) when their value is above 9. The robotic agent has to discriminate between circles and

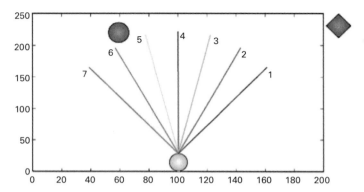

Figure 2.6
Experimental scenario. The agent (gray circle at the bottom) has to catch falling circles and avoid squares in task 1 and catch falling circles with normal and inverted vision in task 2. The robotic agent has seven ray sensors, symmetrically displaced with relation to the central ray in intervals of ±12 radians, and two motors that can move it horizontally. The numbers next to the sensors show the correspondence between the agent's sensory input and the nodes of the network.

squares as they fall from the top of the arena to the bottom (only one object at each trial), where the robotic agent is located. The square's diagonal, the robotic agent's radius, and the circle's radius all measure 15 units. At the beginning of each trial, a circle or a square is dropped from the top of the environment in a random horizontal position within a maximum of 50 units from the robotic agent, and moves vertically with a velocity of 3 units/step. The robotic agent, constrained to move (left and right) in a straight line along the bottom, has to approach the circles and avoid the squares, adjusting its horizontal velocity accordingly.

The second experiment consists of an orientation task. In the same environmental setup, the robotic agent has to adjust its horizontal position to catch falling circles, with both normal and inverted vision. When submitted to visual inversion, sensory readings from an object at the right side of the agent are perceived by the agent's left set of sensors, and vice versa. Therefore, different scenarios can cause similar or identical sensory stimulus to the robotic agent; the agent is required to develop a strategy that can overcome the sensory disruption.

A genetic algorithm is used to determine the parameters of the system: the frequency of each node, $\omega_i \in [0,10]$, the coupling factor $k \in [0,5]$, the input weights $z_i \in [0,3]$, the matrix W with elements in the interval $[-2,2]$, and motor output weight $s \in [0,10]$, resulting in a genotype of length 58 for the tasks studied here. The network's genotype consists of an array of integer variables lying in the range $[0,999]$ (each variable occupies a gene locus), which are mapped to values determined by the range of their respective parameters. For all the experiments reported here, a distributed GA similar to that described in the GasNets section earlier (and in Husbands et al. 1998) was used; the population size was 49, arranged in a 7×7 grid. A generation is defined as 49 breeding events and the evolutionary algorithm runs for a maximum of 150 generations. Two mutation operators are used: the first operator is applied to 20 percent of the genes and produces a change at each locus by an amount within the [−10,+10] range according to a uniform distribution. The second, more disruptive, mutation operator is applied with a probability of 10 percent and is applied to 40 percent of the genotype. A randomly chosen gene locus is replaced with a new value within the [0,999] range in a uniform distribution.

In the first experiment, fitness is evaluated over a set of twenty-eight trials with randomly chosen objects (circles or squares), starting at a uniformly distributed horizontal offset in the interval of ±50 units from the robotic agent. Fitness is defined as the robotic agent's ability to catch circles and avoid squares, and is calculated according to the following function: fitness $= \sum_{i=1}^{N} i f_i / \sum_{i=1}^{N} i$ where f_i is the ith value in a descending ordered vector of evaluation scores for separate trials, and is given by $1 - d_i$, in the case of a circle, or by d_i in the case of a square where d_i is the (normalized) horizontal distance from the robotic agent to the object at the end of the ith trial (when the object reaches the bottom of the environment). Therefore, a robotic agent with good fitness maximizes its distance from squares and minimizes its distance from circles. Note that the form of the fitness function is the same as that of equation 2.6, providing strong pressure for good performance in all trials (generalization).

In the second experiment, fitness is evaluated over a set of twenty trials with normal vision followed by twenty trials with inverted vision. The circles are dropped at a uniformly distributed horizontal offset in the interval of ±50 units from the robotic agent. Fitness for each part of the run is defined as above but considering just the circle-catching scenario. The final fitness is calculated by averaging the fitness obtained under normal and inverted vision.

Results

In the first experiment (catch circles, avoid squares) two network topologies were investigated, a ring topology and a fully connected topology. Both architectures produced similarly good results, providing an existence proof that oscillator phase dynamics can be useful, as part of a situated embodied system, in driving autonomous sensorimotor behavior. The training fitness of the best ring topology individual was 0.96 out of 1.00, and the generalization fitness over 100 random runs was 0.94, which is comparable with the results that are found in the literature (Beer 2003; Di Paolo 2000). For the fully connected topology the training fitness of the best evolved individual was 0.97, and the generalization fitness over 100 runs was 0.92. Analysis of

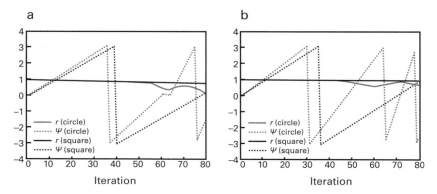

Figure 2.7
Synchronization index (r) and the mean phase (ψ) of the network for the ring topology (a) and
the fully connected topology (b), calculated according to equation 2.8 for best individual evolved
in the first experiment.

successful individuals reveals clearly different dynamics are at play in circle catching
and square avoidance; this is true for both styles of network architecture. Further,
although synchronized states play an important role, unsynchronized and transient
dynamics are also significant (Moioli, Vargas, and Husbands 2010, 2012; Santos, Barandiaran, and Husbands 2012). Figure 2.7 shows the synchronization index plotted
against time for the best individuals from experiment 1. This clearly shows that the
phase dynamics are different for circle catching and square avoidance for both architectures. Square avoidance mainly exploits synchronized dynamics (r for square catching can be seen to always remain very close to 1), whereas circle catching makes
significant use of unsynchronized dynamics (r for circle catching deviates significantly
from 1 after about 40 iterations).

In the second experiment (catching falling circles under normal and inverted
vision), the robotic agent was controlled by the fully connected network architecture,
given its slightly better performance obtained in the previous experiment in the catching circles part of the task. Again agents with very good performance were successfully
evolved. The training fitness for the best evolved individual was 0.94 out of 1.00,
and the generalization fitness over 100 random trails was 0.93. Generalization
performance is illustrated in figure 2.8. Under both conditions, the main adopted
strategy seemed to be to move to one side of the object (in this case, the left side),
where robotic agents with normal visual have their right sensors stimulated while
robotic agents with inverted vision have their left sensors stimulated, and then center
in on the object.

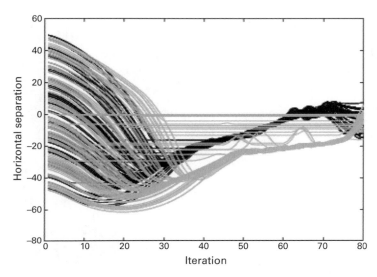

Figure 2.8
The generalization performance of the agent over 100 random trails in experiment 2. The red color is related to the normal vision case and the blue color to the inverted vision scenario. The plot illustrates the value of the horizontal separation of the agent and the object over eighty iterations.

It is possible to see in figure 2.9 that although the strategy for the normal and inverted vision tasks is almost the same, the oscillatory activity of the network and its phase dynamics are quite different, illustrating the multiple roles a single oscillator can have in the network. For example, near iteration 60, under the normal vision condition (upper part of the figure, middle graphic), the frequency of each node is varying and there is no apparent synchronization. Near the same iteration, for the inverted vision case (bottom part of the figure), one can observe two almost synchronized clusters appearing: one formed by 4 nodes, the other formed by 2 nodes and the unsynchronized remaining node oscillating at a much higher frequency. This demonstrates an interesting adaptive mechanism that does not require changes in synaptic strengths but rather works by changing phase and degree of synchronization within neural subgroups that form, break apart, and reconfigure throughout the duration of the behavior. This is another example of the "shifting network" (Husbands et al. 2001) where reconfiguring network dynamics underlies plasticity of behavior.

This analysis again illustrates that evolution is able to readily exploit the rich dynamics the networks are capable of and, importantly, does not rely solely on simple synchronized states in behavior generation. This work is an initial step along a path

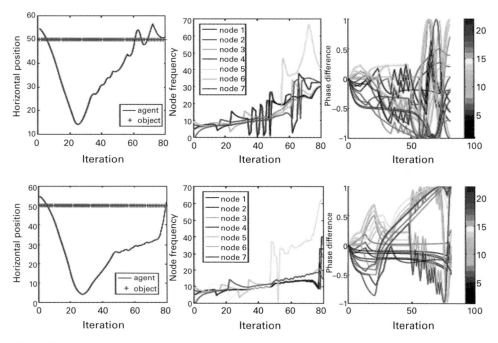

Figure 2.9
Detailed behavior of the agent's internal and external dynamics in experiment 2. The left column illustrates the horizontal coordinate of the agent and the object, the middle one shows the frequency of each node of the network as the task progresses, and the right one presents the phase differences.

that could eventually provide insights into the role of synchronized (and unsynchronized) neuronal dynamics in the generation of cognitive and sensorimotor behaviors. For an extended discussion of related work, including information theoretic analyses of evolved phase-coupled systems, see Moioli, Vargas, and Husbands 2012 and Santos, Barandiaran, and Husbands 2012.

2.3.3 Exploiting Chaotic Dynamics in an Embodied System
This section describes research that is not strictly evolutionary robotics but is closely related and highlights some important emerging topics that are highly relevant to ER and its intersection with neuroscience.

The possibility of exploiting intrinsic chaotic dynamics has recently attracted the attention of both neurobiologists interested in how animals learn to coordinate their limbs (Kelso 1995; Korn and Faure 2003; Mpitsos et al. 1988), for instance in locomotion behaviors, and roboticists striving to develop better, more efficient locomotion systems for articulated autonomous robots (Kuniyoshi and Suzuki 2004; Pitti, Niiyama, and Kuniyoshi 2010; Steingrube et al. 2010). Chaotic dynamics emerging spontaneously from interactions between neural circuitry and bodies and environments can be used to power a kind of search process as an embodied system explores its own possible motor behaviors. However, to date it has not been clear how to harness chaos in a general goal-directed way such that desired adaptive sensorimotor behaviors can be explored, captured, and learned. In this section we briefly present a general and fully dynamic embodied neural system, which exploits chaotic search through adaptive bifurcation, for the real-time goal-directed exploration and learning of the possible locomotion patterns of an articulated robot of an arbitrary morphology in an unknown environment. Our results show that the novel neuro-robotic system is able to create and learn a number of emergent locomotion behaviors for a wide range of body configurations and physical environments, and can readapt in real time after sustaining damage. For further details of the methods, see Shim and Husbands 2010 and 2012.

Properly coordinated rhythmic movements for locomotion are ubiquitous in animals. Biological locomotor systems (usually involving coordinated limb movements) evolved to be highly adaptable, dexterous, and energy efficient. Consequently they are a major source of inspiration when designing robot locomotion systems. Most biological locomotor systems involve neural networks acting as central pattern generators (CPGs), which are responsible for producing the basic rhythmic patterns for the oscillatory movement of limbs. Understanding the subtleties of operation of such networks, and how to design artificial versions for robotic applications, are ongoing challenges.

Most approaches to designing CPG-based robotic locomotor systems have relied on optimization and search methods, including evolutionary algorithms and other stochastic methods, to find a suitable configuration of system variables—including the ER approaches to locomotion described earlier. These methods can be computationally expensive and often require a priori knowledge of the robot body and environment. Hence there are still many open issues in how to deal with unknown environments and adaptation to arbitrary or changed (e.g., damaged) body conditions in the most general and efficient way.

In robotics, a greater appreciation of the importance of framing behavior in terms of brain-body-environment interactions has led to efforts to exploit various ready-made functionalities provided by the physical properties of an embodied system. A recent strand of work has built on the growing body of observations of intrinsic chaotic dynamics in nervous systems to suggest that such dynamics can underpin crucial periods in animal development when brain-body-environment dynamics are explored in a spontaneous way as part of the process of acquiring motor skills. Recent robotics studies have demonstrated that chaotic neural networks can indeed power the self-exploration of brain-body-environment dynamics in an embodied system, discovering stable patterns that can be incorporated into motor behaviors (Pitti, Niiyama, and Kuniyoshi 2010; Kuniyoshi and Sangawa 2006). In the work outlined here we significantly generalize and extend this previous research to demonstrate how chaotic neural dynamics can be harnessed to develop a kind of system not seen in previous models: one where intrinsic neural dynamics can be used to autonomously explore, capture, and learn whole goal-directed sensorimotor behaviors in an embodied system without recourse to external monitoring, evaluation, or training methods. We introduce a general online and fully dynamic neural process for the exploration and learning of possible locomotion patterns for articulated robots of an arbitrary morphology in unknown physical environments. Goal-directed exploration is achieved using chaotic search while discovered patterns are memorized and sustained by adaptive changes to the wirings of chaotic neural oscillators that form the basis of the neural architecture. As well as having direct application in robotics, this work has potential implications for neurobiology.

Conventional optimization and search strategies generally use random perturbations of the system variables to search the space of possible solutions. We have developed a method that uses the intrinsic chaotic dynamics of the system to naturally power a search process without the need for external sources of noise (Shim and Husbands 2010, 2012). We employ the concept of chaotic mode transition with external feedback (Davis 1990), which exploits the intrinsic chaoticity of a system orbit as a perturbation force to explore multiple synchronized states of the system, and stabilizes the orbit by decreasing its chaoticity according to a feedback signal that evaluates the behavior. An evaluation signal that measures how well the locomotion behavior of the system matches the desired criteria (e.g., locomote as fast as possible) is used to control a bifurcation parameter that alters the chaoticity of the system. During exploration, the bifurcation parameter continuously drives the system between stable and chaotic regimes. If the performance reaches the desired level, the bifurcation parameter decreases to zero and the system stabilizes. A learning process that acts in tandem with the chaotic exploration captures and memorizes these high-performing motor patterns.

The overall architecture of the system is illustrated in figure 2.10; the neural architecture generalizes and extends that presented in Kuniyoshi and Sangawa 2006, which is inspired by the organization of spinobulbar units in the vertebrate spinal cord and medulla oblongata (the lower part of the brainstem that deals with autonomic, rhythmic, involuntary functions). Each joint in an articulated robot is connected to a motor unit comprising a pair of central pattern generator (CPG) neurons that receive sensory input. The neurons produce motor outputs for an antagonistic muscle pair that control the movement of the joint. Each CPG neuron is modeled as an extended Bonhöffer van der Pol (BVP) oscillator (FitzHugh 1961; Asai et al. 2003), which can be viewed as a simplification of the full Hodgkin Huxley neural model (Hodgkin and Huxley 1952). The CPG neurons are all connected to each other but these connections are initially

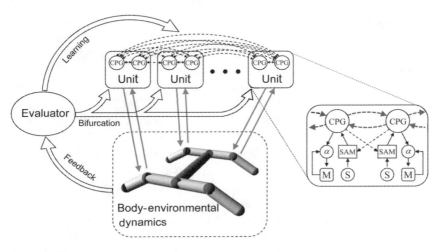

Figure 2.10

An overview of the integrated exploration and learning scheme. Each robot joint has a dedicated motor unit comprising oscillator-based central pattern generator neurons (CPG) with sensory input (S) and motor output (M). Connections between the oscillators are initially inactive but they are weakly coupled through the body and environment. An evaluation feedback signal controls a global bifurcation parameter that alters the chaoticity of the CPGs. The chaotic dynamics of the neutron-body-environment system drive a search process that finds motor patterns that perform well according to the evaluation criteria. As the system stabilizes on a high-performing pattern the bifurcation parameter reduces to zero and the connections between the oscillators become active, their weights being set by a learning procedure that is smoothly linked to the chaotic exploration process. The learning process further stabilizes, captures, and memorizes the motor patterns. Sensory input undergoes homeostatic adaptation as it passes through a sensor adaptation module (SAM). This enhances the synchronicity between the neural and physical system, thus allowing the neural system to cope with an arbitrary robotic system.

made inactive. The CPG neurons receive sensory signals that integrate information from the body-environment interaction dynamics experienced by the system (e.g., from force and position/angle sensors). Hence, while the direct connections are inactive, any coupling between the oscillators will be indirect via bodily and environmental interactions. The network of oscillators, coupled through physical embodiment, has multiple synchronized states (modes) that reflect the body schema and its interactions with the environment, each of which can be regarded as a potential candidate for "meaningful" motor behavior. The exploration process, powered by adaptive bifurcation through the feedback evaluation signal, allows the system to become entrained in these modes, one at a time, until one is found that is sufficiently stable and high performing for the bifurcation parameter to reduce to zero and the system to fully stabilize. As the system stabilizes, the connections between oscillators are dynamically activated using an adaptive synchronization-learning scheme. In this way the wiring between the oscillators is changed in order to capture and maintain the high-performing motor pattern. The learning rule is also controlled by the bifurcation parameter and is set up such that the connections between the oscillators are effectively zero (inactive) during the exploration process but gradually increase (become active) as the system nears stability (see Shim and Husbands 2012 for full details). Thus, exploration and learning are merged as a continuous dynamical process such that the desired motor behavior is spontaneously explored, discovered, and memorized in a coherent way. If the performance drops, for instance following a change in the environment or damage to the body, the system will automatically return to the exploration phase until a new stable high-performing motor pattern is discovered. The overall process has some conceptual similarities with Ashby's idea of the ultrastable system (Ashby 1952), although, not surprisingly given how long ago he conceived it, Ashby only envisaged simple stochastic perturbations. The method can be interpreted as a continuous and deterministic version of trial-and-error search that exploits the intrinsic chaotic behavior of the system.

The pair of CPG neurons in each motor unit (labeled l and r) operates according to the following coupled differential equations:

$$\tau \dot{x}_l = c\left(x_l - \frac{x_l^3}{3} - y_l + z_1\right) + \delta\left(H_l\left(s_l\right) - x_l\right) + F_l^x \tag{2.10}$$

$$\tau \dot{y}_l = \frac{1}{c}\left(x_l - by_l + a\right) + \varepsilon\left(H_l\left(s_l\right)\right) + F_l^y \tag{2.11}$$

$$\tau \dot{x}_r = c\left(x_r - \frac{x_r^3}{3} - y_r + z_2\right) + \delta\left(H_r\left(s_r\right) - x_r\right) + F_r^x \tag{2.12}$$

$$\tau \dot{y}_r = \frac{1}{c}\left(x_r - by_r + a\right) + \varepsilon\left(H_r\left(s_r\right)\right) + F_r^y \tag{2.13}$$

Where τ is a time constant, and $a = 0.7$, $b = 0.675$, $c = 1.75$ are the fixed parameters of the oscillator. Each consecutive pair in the set of 2N oscillators are sequentially allocated to each motor unit as $l = 2m - 1$ and $r = 2m$. $\delta = 0.013$ and $\varepsilon = 0.022$ are the coupling strengths for afferent input $H(s)$ that is a function of raw sensor output s, processed by the sensor adaptation module (SAM)—see figure 2.10. F_i^j is a coupling term between oscillators and is subject to the learning process. z_1 and z_2 are the control parameters for adjusting the chaoticity of the motor unit. Their difference ($\mu = z_2 - z_1$) changes identically in all motor units, and acts as the global bifurcation parameter. In the stable regime where the two control parameters are symmetric, it had been found (Asai et al. 2003) that the two coupled BVP equations exhibit bistable phase locking of their oscillations in a parameter range of $0.6 < z_1 = z_2 < 0.88$. From the observation of a number of experiments on the oscillator dynamics, we chose to fix $z_2 = 0.73$ and to vary z_1 in order to ensure a higher probability of multistability of the system in its stable regime. For further details, see Shim and Husbands 2012.

The evaluation signal is determined by a ratio of the actual performance (e.g., forward speed) to the desired performance. Where the desired behavior is forward locomotion, the evaluation signal, E, is measured according to the following equations, where $\boldsymbol{v}$ is the robot velocity vector and τ_E is a time constant. Using this leaky integrator equation means the velocity is continuously averaged over a time window, thus eradicating gyrations and oscillations.

$$E(t) = \bar{\boldsymbol{v}}, \ \tau_E \frac{d\bar{\boldsymbol{v}}}{dt} = -\bar{\boldsymbol{v}} + \boldsymbol{v} \tag{2.14}$$

The time course of the bifurcation parameter, μ, is tied to the evaluation signal using the following equations.

$$\tau_\mu \frac{d\mu}{dt} = -\mu + \mu_c G\left(\frac{E}{E_d}\right), \ G(x) = \frac{1}{(1 + e^{16x-8})} \tag{2.15}$$

Where τ_μ and μ_c are constants and E_d is the desired performance; $G(x)$ implements a decreasing sigmoid function that maps monotonically from (0, 1) to (1, 0) shaped so that the boundary value at $x=1$ and its derivative become almost 0 so as to make the function smoothly vanish to zero to facilitate gradual stabilization. Since the method is intended for use in the most general case, where the robotic system is arbitrary, we do not have prior knowledge of what level of performance it can achieve. Using the concept of a goal-setting strategy (Barlas and Yasarcan 2006), the dynamics of the desired performance are modeled as a temporal average of the actual performance, such that the expectation of a desired goal is influenced by the history of the actual performance experienced as described by the following equation.

$$\tau_d \frac{dE_d}{dt} = -E_d + E \tag{2.16}$$

The sensor signal fed to a CPG neuron undergoes homeostatic adaptation as it passes through a sensor adaptation module (SAM) before reaching the neuron (see figure 2.10). The SAMs were introduced because by adjusting the waveforms of input signals to be close to those of the neural activities, the synchronicity between the neural and physical system was enhanced thus allowing the neural system to cope with an arbitrary robotic system. This regulation also results in a diversification of output behaviors, increasing the scale of the search process.

The chaotic exploration and learning system was evaluated by using it to control a range of realistically simulated articulated robots that were required to locomote in an effective way. Successful experiments with a swimming robot and a variety of walking robots of differing morphologies demonstrated that the framework is highly general

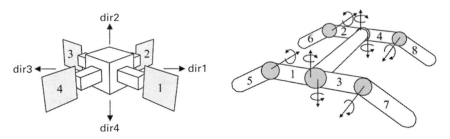

Figure 2.11
Robotic simulation models of a 4-Fin Swimmer (2D movement) and a quadruped (3D movement) used to evaluate the method.

(Shim and Husbands 2012). In each case a range of stable locomotion behaviors were discovered and learned. It was also shown that the robots can readily readapt after damage or other changes.

The seamless interaction between the exploration and learning processes results in a system that can be thought of as continually self-monitoring in order to maintain an appropriate level of motor function. As well as being an effective means of developing robotic controllers, the method has more general implications for truly autonomous artificial systems, which must maintain their integrity on several levels, including behavioral. Because of its strong biological inspiration it also serves as an indication of the kinds of processes that may be operating in natural nervous systems.

Figure 2.11 shows two of the robot simulations used to demonstrate the system. Figure 2.12 shows a typical swimming motion discovered and captured for the 4-Fin Swimmer. It is an efficient frog-like motion. Figure 2.13 illustrates how the method allows real-time adaptation. The robot suffered damage where the length of the fin on arm 4 was reduced by 90 percent. Its performance dropped almost immediately to zero so the exploration process automatically kicked in. Very quickly it discovered and stabilized one of the high-performing transient patterns, needing only a few trials for the oscillator learning mechanism to completely stabilize the system once more into a new high-performance swimming motion that is able to compensate for the damage.

Figure 2.14 illustrates the generality of the method by showing how it quickly discovers, captures, and stabilizes high-performing walking behavior in a quadruped robot model. For details of applications of the method to other robots and body morphologies, see Shim and Husbands 2012.

Just as incorporating various forms of neural plasticity into ER has proved very fruitful (Floreano and Urzelai 2000; Husbands et al. 1998; Di Paolo 2003), the integration of chaotic dynamics, as outlined in this section, may result in a powerful hybrid

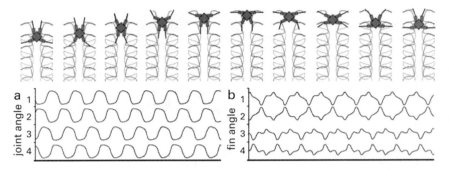

Figure 2.12
Upper: snapshots of a straight swimming (ST dir3) behavior of the 4-Fin Swimmer developed by the exploration and capture method. Images were taken every 1/10 gait cycle. The tip trajectories of the fore (fin 3,4:black) and rear (fin1,2:grey) fins are shown. Lower: (a) joint angles and (b) fin-bending angles of the behavior. Each segment along the vertical axis indicates the range [−1,1] rad.

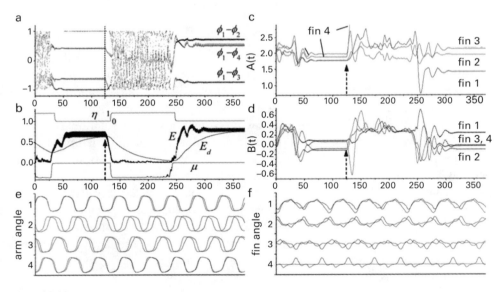

Figure 2.13
Real-time recovery after a radical change to the body of the swimmer (damage). Dashed lines and arrows indicate the time of damage, when the length of fin 4 is decreased to 1/10 of its original length. The sensor gain of (damaged) fin 4 ($A(t) \approx 5.0$) in (c) was truncated for a better view of the other gain plots. (e) and (f) show the joint angles and the fin angles respectively, where the undamaged motion and the readapted motion are superposed. The fiducial point for the super-posed plots was set to the starting point of arm angle 1 in (e).

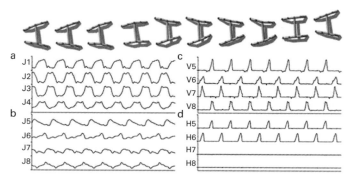

Figure 2.14
An example of a quadruped gait captured by the exploration-learning process. Snapshots were taken every 1/10 gait cycle. (a) and (b) show the joint angles of limbs. (c) shows the horizontal speeds of each foot (the tips of limbs 5–8) in the direction of locomotion. (d) shows the height of each foot from the ground. The two rear feet (V7, V8, H7, H8) show stick-and-slip movements on the ground under Coulomb friction. The range of each plot is as follows; J1–J8: [−1.0,1.0]rad, V5–V8: [0.0,2.0]m/s, H5–H8: [0.0,0.08]m.

method. There are a number of parameters in the chaotic method that need to be set by hand. ER methods may be able find better setting for combinations of these parameters in order to further increase the efficiency of the method. But a potentially more powerful hybridization of the chaotic search and learning method with ER will be to allow the exploration of more complex architectures within an evolutionary framework. This may involve the integration of spinobulbar motor units within a more complex overall architecture, for instance to allow a variety of behaviors, or the development of alternative architectures for the motor units themselves.

The chaotic search approach has some commonalities with information-driven approaches to ER (Delarboulas, Schoenauer, and Sebag 2012) and to self-organization (Burfoot, Lungarella, and Kuniyoshi 2008; Zahedi, Ay, and Der 2010). The former make use of information theoretic fitness measures (e.g., entropy of a stream of sensor and motor readings) to encourage, for instance, the behavioral diversity of controllers that will continually seek to explore.

2.4   Discussion and Prospects

Although great progress has been made in autonomous robotics over the past few decades, and techniques similar to those discussed in this chapter have played their part in some of the ever-proliferating mobile robots we now see in the home (e.g., autonomous vacuum cleaners and toys), or in areas such as planetary exploration, security, or military applications, many challenges remain.

It is now possible to produce autonomous robots that behave in a robust and reliable way in real environments, engaging in real tasks in real time. However, the behaviors involved are still relatively simple. Progress has been slow toward more sophisticated tasks such as learning what to focus attention on in a complex environment, coordinating many conflicting goals and drives, interacting with other robots in complex group behaviors, learning to communicate in a sophisticated way about the task at hand, and so on. Perhaps this should not be at all surprising. One lesson that most neuroscientists have understood for many decades, but which has often been overlooked in AI, is that the generation of intelligent embodied behavior is an extremely complicated problem. However, progress is being made and there are many promising lines of research. It is likely that directions involving artificial neural systems and other biologically inspired methods will become even more important as attempts to tackle these hard problems gather momentum. ER methods will surely play a role, either as a standalone direction or in concert with other approaches. One direction, mentioned in several examples described earlier, that is likely to become increasingly important is the continued dismantling of the line between brain and body that has traditionally been present in studies of both natural and artificial intelligence. The tighter integration of artificial bodies and brains at many different levels, as Pollack suggests (chapter 12, this volume), is an intriguing possibility that would probably require advances in evolutionary developmental systems—something that could usher in an exciting new direction in ER research.

There are a number of potentially important emerging fields that may have a radical impact in the decades to come. These include developments in interfacing digital electronics to neural tissue. The most frequent motivation for such work is to allow improved prosthetics to be directly controlled by the nervous system. This points to the possibility of an increased merging of robotic technology with human bodies—something that a number of people have reflected on recently (e.g., Brooks 2002) and that the work of Stelarc, the radical performance artist, has long explored (Smith 2007). A related area involves attempting to harness the sophisticated nonlinear adaptive properties of cultured (real) neural networks to create hybrid machines (DeMarse et al. 2001), pointing toward the possibility of robots that include biological matter in their control systems—a development that would echo the imagined landscapes of dozens of sci-fi books and movies. It is possible that in the long run that kind of approach may prove more powerful than attempting to understand biological systems in sufficient detail to be able to abstract general mechanisms underlying the generation of intelligent behavior. However, such research is at an extremely early stage, so we cannot yet properly assess its potential. One approach that has been considered recently is the use of artificial selection to shape cultured networks toward some behavior end (Bull 2004). This is very difficult with current technology, but advances in high-density multielectrode array hardware, allied with chemical perfusion systems, might allow sufficiently powerful and repeatable manipulation of neural networks to make this a viable approach.

There is little doubt that many powerful biological neural mechanisms have not yet been discovered. One intriguing hypothesis is that one of the forms of plasticity on which the brain relies is itself a form of evolution via natural selection acting within neural tissue (Fernando, Karishma, and Szathmáry 2008; Fernando, Szathmary, and Husbands 2012). The units of selection in this case are activity and connection patterns that are copied between groups of neurons. Irrespective of whether or not it occurs in nature (and it might), this kind of mechanism could be employed in a whole new kind of evolutionary robotics.

The field of robotics has massively expanded since the days when cumbersome industrial arms dominated; it is now quite possible that in the not too distant future robots will become as widespread and as common as computers are now. If such a technological revolution comes to pass, it is highly likely that artificial neural systems will play an important part as there will be greater demands for robust, reliable adaptation and learning, as well as sophisticated pattern recognition and sensory processing—all areas in which neural systems have great potential, especially within the context of behaving embodied systems. This in turn means that there are great opportunities for the kinds of interactions between ER and neuroscience described in this chapter.

References

AB. 2009. *Adaptive Behavior* 17 (4). Special issue on robotic and agent modeling of adaptive behavior, based around responses to Webb.

Abbott, L. F., and T. B. Kepler. 1990. Model neurons: From Hodgkin-Huxley to Hopfield. In *Statistical Mechanics of Neural Networks*, ed. L. Garrido, 5–18. Berlin: Springer.

Acebron, J., L. Bonilla, C. PerezVicente, F. Ritort, and R. Spigler. 2005. The Kuramoto model: A simple paradigm for synchronization phenomena. *Reviews of Modern Physics* 77 (1): 137–185.

Asai, Y., T. Nomura, S. Sato, A. Tamaki, Y. Matsuo, I. Mizukura, and K. Abe. 2003. A coupled oscillator model of disordered interlimb coordination in patients with Parkinson's disease. *Biological Cybernetics* 88:152–162.

Ashby, W. R. 1952. *Design for a Brain*. London: Chapman and Hall.

Baranano, D., C. Ferris, and S. Snyder. 2001. Atypical neural messengers. *Trends in Neurosciences* 24 (2): 99–106.

Barlas, Y., and H. Yasarcan. 2006. Goal setting, evaluation, learning and revision: A dynamic modeling approach. *Evaluation and Program Planning* 29 (1): 79–87.

Barnett, L. C., and A. K. Seth. 2011. Behaviour of Granger causality under filtering: Theoretical invariance and practical application. *Journal of Neuroscience Methods* 201:404–419.

Beer, R. 1995. A dynamical systems perspective on environment agent interactions. *Artificial Intelligence* 72:173–215.

Beer, R. 2003. The dynamics of active categorical perception in an evolved model agent. *Adaptive Behavior* 11 (4): 209–243.

Beer, R., and J. Gallagher. 1992. Evolving dynamical neural networks for adaptive behaviour. *Adaptive Behavior* 1:94–110.

Beer, R. D., H. J. Chiel, and L. S. Sterling. 1989. Heterogeneous neural networks for adaptive behavior in dynamic environments. In *Neural Information Processing Systems 1*, ed. D. Touretzky, 577–585. San Francisco: Morgan Kauffman.

Beer, R. D., R. D. Quinn, H. J. Chiel, and R. E. Ritzmann. 1997. Biologically-inspired approaches to robotics. *Communications of the ACM* 40 (3): 30–38.

Berger, H. 1929. Uber das elektrenkephalogramm des menschen. *Archiv für Psychiatrie und Nervenkrankheiten* 87:527–570.

Brooks, R. 1991. Intelligence without representation. *Artificial Intelligence* 47:139–159.

Brooks, R. A. 2002. *Flesh and Machines: How Robots Will Change Us*. New York: Pantheon Books.

Bull, Larry. 2004. Personal communication.

Bullock, T., M. Bennett, D. Johnston, R. Josephson, E. Marder, and R. Fields. 2005. The neuron doctrine, redux. *Science* 310 (5749): 791.

Burfoot, D., M. Lungarella, and Y. Kuniyoshi. 2008. Toward a theory of embodied statistical learning. In *Proceedings of the 10th International Conference on the Simulation of Adaptive Behavior (SAB '08)*, LNCS 5040, ed. M. Asada, J. Hallam, J.-A. Meyer, and J. Tani, 270–279. Berlin: Springer.

Buzsaki, G. 2006. *Rhythms of the Brain*. Oxford: Oxford University Press.

Clark, A. 1999. An embodied cognitive science? *Trends in Cognitive Sciences* 3 (9): 345–351.

Cliff, D., I. Harvey, and P. Husbands. 1993. Explorations in evolutionary robotics. *Adaptive Behavior* 2:73–110.

Collins, R., and D. Jefferson. 1991. Selection in massively parallel genetic algorithms. In *Proceedings of the 4th International Conferenceon Genetic Algorithms*, ed. R. Belew, 249–256. San Francisco: Morgan Kaufmann.

Cumin, D., and C. Unsworth. 2007. Generalizing the Kuramoto model for the study of neuronal synchronization in the brain. *Physica D. Nonlinear Phenomena* 226 (2): 181–196.

Dale, K., and P. Husbands. 2010. The Evolution of reaction-diffusion controllers for minimally cognitive agents. *Artificial Life* 16 (1): 1–19.

Davis, G. 2006. Homeostatic control of neural activity: From phenomenology to molecular design. *Annual Review of Neuroscience* 29:307–323.

Davis, P. 1990. Application of optical chaos to temporal pattem search in a nonlinear optical resonator. *Japanese Journal of Applied Physics* 29:L1238–L1240.

Dawson, T., and S. Snyder. 1994. Gases as biological messengers: Nitric oxide and carbon monoxide in the brain. *Journal of Neuroscience* 14 (9): 5147–5159.

Dayan, P., and L. F. Abbott. 2001. *Theoretical Neuroscience: Computational and Mathematical Modeling of Neural Systems*. Cambridge, MA: MIT Press.

de Garis, H. 1990. Genetic programming: Evolution of time dependent neural network modules which teach a pair of stick legs to walk. In *Proceedings of the 9th European Conference on Artificial Intelligence*, ed. L. Aiello, 204–206. London: Pitman.

Delarboulas, P., M. Schoenauer, and M. Sebag. 2010. Open-ended evolutionary robotics: An information theoretic approach. *In Proceedings of 11th International Conference on Parallel Problem Solving from Nature* (PPSN XI), LNCS 6238, ed. R. Schaefer, C. Cotta, J. Kołodziej, G. Rudolph, 334–343. Berlin: Springer.

DeMarse, T., D. Wagenaar, A. Blau, and S. Potter. 2001. The neurally controlled animat: Biological brains acting with simulated bodies. *Autonomous Robots* 11:305–310.

Di Paolo, E. 2000. Homeostatic adaptation to inversion of the visual field and other sensorimotor disruptions. In *From Animals to Animats 6, SAB 2000*, ed. J.-A. Meyer, A. Berthoz, D. Floreano, H. Roitblat, and S. Wilson, 440–449. Cambridge, MA: MIT Press.

Di Paolo, E. 2003. Evolving spike-timing-dependent plasticity for single-trial learning in robots. *Philosophical Transactions of the Royal Society of London. Series A, Mathematical and Physical Sciences* 361:2299–2319.

Doya, K. 2002. Metalearning and neuromodulation. *Neural Networks* 15:495–506.

Ekeberg, O. 1993. A combined neuronal and mechanical model of fish swimming. *Biological Cybernetics* 69:363–374.

Ekeberg, O., A. Lansner, and S. Grillner. 1995. The neural control of fish swimming studied through numerical simulations. *Adaptive Behavior* 3 (4): 363–384.

El-Nashar, H., and H. Cerdeira. 2009. Determination of the critical coupling for oscillators in a ring. *Chaos* 19 (3): 033127.

Engel, A., P. Fries, and W. Singer. 2001. Dynamic predictions: Oscillations and synchrony in top-down processing. *Nature Reviews Neuroscience* 2 (10): 704–716.

Fernando, C., K. K. Karishma, and E. Szathmáry. 2008. Copying and evolution of neuronal topology. *PLoS ONE* 3 (11): e3775. doi:10.1371/journal.pone.0003775.

Fernando, C., E. Szathmary, and P. Husbands. 2012. Selectionist and evolutionary approaches to brain function: A critical appraisal. *Frontiers in Computational Neuroscience* 6:24. doi:10.3389/fncom.2012.00024.

Ferree, T. C., and S. R. Lockery. 1999. Computational rules for chemotaxis in the nematode C. elegans. *Journal of Computational Neuroscience* 6:263–277.

FitzHugh, R. 1961. Impulses and physiological states in theoretical models of nerve membrane. *Biophysical Journal* 1:445–466.

Floreano, D., P. Husbands, and S. Nolfi. 2008. Evolutionary robotics. In *Springer Handbook of Robotics*, ed. B. Siciliano and O. Khatib, 1423–1451. Berlin: Springer.

Floreano, D., and C. Mattiussi. 2001. Evolution of spiking neural controllers for autonomous vision-based robots. In *Evolutionary Robotics: From Intelligent Robotics to Artificial Life*, ed. T. Gomi, 38–61. Tokyo: Springer-Verlag.

Floreano, D., and F. Mondada. 1994. Automatic creation of an autonomous agent: Genetic evolution of a neural-network driven robot. In *From Animals to Animats 3: Proceedings of the Third International Conference on Simulation of Adaptive Behavior*, ed. D. T. Cliff, P. Husbands, J.-A. Meyer, and S. Wilson, 402–410. Cambridge, MA: Bradford Books/MIT Press.

Floreano, D., and J. Urzelai. 2000. Evolutionary robots with on-line self-organization and behavioral fitness. *Neural Networks* 13 (4–5): 431–443.

Frigg, R., and S. Hartmann. 2008. Models in science. *The Stanford Encyclopedia of Philosophy* (Fall 2008 edition), ed. Edward N. Zalta. http://plato.stanford.edu/archives/fall2008/entries/models-science/> (accessed January 2010).

Friston, K. 2000. The labile brain. i. Neuronal transients and nonlinear coupling. *Philosophical Transactions of the Royal Society of London. Series B, Biological Sciences* 355:215–236.

Gally, J., P. Montague, G. Reeke, and G. Edelman. 1990. The NO hypothesis: Possible effects of a short-lived, rapidly diffusible signal in the development and function of the nervous system. *Proceedings of the National Academy of Sciences of the United States of America* 87:3547–3551.

Gerken, W. C., L. K. Purvis, and R. J. Butera. 2005. Use of a genetic algorithm for neuron model specification. In *Proceedings of the 2nd International IEEE EMBS Conference on Neural Engineering*, 304–306. Los Alamitos, CA: IEEE Press.

Granger, C. W. J. 1969. Investigating causal relations by econometric models and cross-spectral methods. *Econometrica* 37:424–438.

Grillner, S., P. Wallen, and L. Brodin. 1991. Neuronal network generating locomotor behavior in lamprey: Circuitry, transmitters, membrane properties, and simulation. *Annual Review of Neuroscience* 14:169–199.

Gurkiewicz, M., and A. Korngreen. 2007. A numerical approach to ion channel modelling using whole-cell voltage-clamp recordings and a genetic algorithm. *PLoS Computational Biology* 3:e169.

Hallam, J., and A. Ijspeert. 2003. Using evolutionary methods to parameterize neural models: A study of the lamprey central pattern generator. In *Biologically Inspired Robot Behavior Engineering*, ed. R. J. Duro, J. Santos, and M. Grana, 119–142. Berlin: Springer.

Hartmann, S. 1995. Models as a tool for theory construction: Some strategies of preliminary physics. In *Theories and Models in Scientific Process*, Poznan Studies in the Philosophy of Science and the Humanities 44, ed. W. Herfel, et al., 49–67. Amsterdam: Rodopi.

Harvey, I., E. Di Paolo, R. Wood, M. Quinn, and E. Tuci. 2005. Evolutionary robotics: A new scientific tool for studying cognition. *Artificial Life* 11 (1–2): 79–98.

Harvey, I., P. Husbands, and D. Cliff. 1993. Issues in evolutionary robotics. In *From Animals to Animats 2: Proceedings of the Second International Conference on the Simulation of Adaptive Behaviour*, ed. J.-A. Meyer, H. Roitblat, and S. Wilson, 364–373. Cambridge, MA: MIT Press.

Harvey, I., P. Husbands, and D. Cliff. 1994. Seeing the light: Artificial evolution, real vision. In *From Animals to Animats 3: Proceedings of the Third International Conference on the Simulation of Adaptive Behaviour (SAB '94)*, ed. D. T. Cliff, P. Husbands, J.-A. Meyer, and S. Wilson, 392–401. Cambridge, MA: MIT Press.

Held, R., and A. Hein. 1963. Movement-produced stimulation in the development of visually guided behavior. *Journal of Comparative and Physiological Psychology* 56 (5): 872–876.

Hodgkin, A., and A. Huxley. 1952. A quantitative description of membrane current and its application to conduction and excitation in nerve. *Journal of Physiology* 117:500–544.

Husbands, P. 1992. An ecosystems model for integrated production planning. *International Journal of Computer Integrated Manufacturing* 6 (1&2): 74–86.

Husbands, P. 1998. Evolving robot behaviours with diffusing gas networks. In *Evolutionary Robotics: First European Workshop, EvoRobot98*, Lecture Notes in Computer Science 1468, ed. P. Husbands and J.-A. Meyer, 71–86. Berlin: Springer.

Husbands, P., and I. Harvey. 1992. Evolution versus design: Controlling autonomous mobile robots. In *Proceedings of the 3rd Annual Conference on Artificial Intelligence, Simulation and Planning in High Autonomy Systems*, 139–146. Los Alamitos, CA: IEEE Computer Society Press.

Husbands, P., I. Harvey, D. Cliff, and G. Miller. 1997. Artificial evolution: A new path for AI? *Brain and Cognition* 34:130–159.

Husbands, P., A. Philippides, T. Smith, and M. O'Shea. 2001. The shifting network: Volume signalling in real and robot nervous systems. In *Proceedings of ECAL2001*, LNAI 2159, ed. J. Kelemen and P. Sosik, 23–37. Berlin: Springer.

Husbands, P., A. Philippides, P. Vargas, C. Buckley, P. Fine, E. Di Paolo, and M. O'Shea. 2010. Spatial, temporal and modulatory factors affecting GasNet evolvability in a visually guided robotics task. *Complexity* 16 (2): 35–44.

Husbands, P., T. Smith, N. Jakobi, and M. O'Shea. 1998. Better living through chemistry: Evolving GasNets for robot control. *Connection Science* 10 (3–4): 185–210.

Ijspeert, A., A. Crespi, D. Ryczko, and J. Cabelguen. 2005. From swimming to walking with a salamander robot driven by a spinal cord model. *Science* 315 (5817): 1416–1420.

Ijspeert, A., J. Hallam, and D. Willshaw. 1999. Evolving swimming controllers for a simulated lamprey with inspiration from neurobiology. *Adaptive Behavior* 7 (2): 151–172.

Izhikevich, E. 1999. Weakly pulse-coupled oscillators, FM interactions, synchronization, and oscillatory associative memory. *IEEE Transactions on Neural Networks* 10 (3): 508–526.

Izhikevich, E. 2007. *Dynamical Systems in Neuroscience: The Geometry of Excitability and Bursting.* Cambridge, MA: MIT Press.

Izquierdo, E. 2006. The dynamics of learning behaviour: A situated, embodied, and dynamical systems approach. PhD thesis, Centre for Computational Neuroscience and Robotics, University of Sussex.

Izquierdo, E. J., and S. R. Lockery. 2010. Evolution and analysis of minimal neural circuits for klinotaxis in Caenorhabditis elegans. *Journal of Neuroscience* 30:12908–12917.

Jakobi, N. 1998. Running across the reality gap: Octopod locomotion evolved in a minimal simulation. In *Evolutionary Robotics: First European Workshop, EvoRobot98*, ed. P. Husbands and J.-A. Meyer, 39–58. Berlin: Springer.

Katz, P., ed. 1999. *Beyond Neurotransmission: Neuromodulation and Its Importance for Information Processing.* Oxford: Oxford University Press.

Kelso, J. A. S. 1995. *Dynamic Patterns: The Self-Organization of Brain and Behavior.* Cambridge, MA: MIT Press.

Kodjabachian, J., and J.-A. Meyer. 1998. Evolution and development of neural networks controlling locomotion, gradient following and obstacle avoidance in artificial insects. *IEEE Transactions on Neural Networks* 9:796–812.

Konig, P., A. Engel, and W. Singer. 1996. Integrator or coincidence detector? The role of the cortical neuron revisited. *Trends in Neurosciences* 19:130–137.

Korn, H., and P. Faure. 2003. Is there chaos in the brain? II. Experimental evidence and related models. *Comptes Rendus Biologies* 326:787–840.

Kuniyoshi, Y., and S. Sangawa. 2006. Early motor development from partially ordered neural-body dynamics: Experiments with a cortico-spinal-musculo-skeletal model. *Biological Cybernetics* 95:589–605.

Kuniyoshi, Y., and S. Suzuki. 2004. Dynamic emergence and adaptation of behaviour through embodiment as coupled chaotic field. In *Proceedings of the IEEE International Conference on Intelligent Robots and Systems*, 2042–2049. Los Alamitos, CA: IEEE Press.

Kuramoto, Y. 1984. *Chemical Oscillation, Waves, and Turbulence.* New York: Springer.

Kuyucu, T., I. Tanev, and K. Shimohara. 2012. Incremental evolution of fast moving and sensing simulated snake-like robot with multiobjective GP and strongly-typed crossover. *Memetic Computing* 4 (3): 183–200.

Li, Z., and J. Hopfield. 1989. Modeling the olfactory bulb and its neural oscillatory processings. *Biological Cybernetics* 61 (5): 379–392.

McHale, G., and P. Husbands. 2004. GasNets and other evolvable neural networks applied to bipedal locomotion. In *From Animals to Animats 8: Proceedings of the Eighth International Conference on Simulation of Adaptive Behavior (SAB '2004)*, ed. S. Schaal et al., 163–172. Cambridge, MA: MIT Press.

Moioli, R., P. Vargas, and P. Husbands. 2010. Exploring the Kuramoto model of coupled oscillators in minimally cognitive evolutionary robotics tasks. In *Proceedings of the IEEE Congress on Evolutionary Compututation 2010 (CEC '10)*. Los Alamitos, CA: IEEE Press.

Moioli, R., P. Vargas, and P. Husbands. 2012. Synchronisation effects on the behavioural performance and information dynamics of a simulated minimally cognitive robotic agent. *Biological Cybernetics* 106 (6–7): 407–427.

Monteiro, L., N. Canto, J. Orsatti, and F. Piqueira. 2003. Global and partial synchronism in phase-locked loop networks. *IEEE Transactions on Neural Networks* 14 (6): 1572–1575.

Mpitsos, G. J., R. M. Burton, H. C. Creech, and S. O. Soinila. 1988. Evidence for chaos in spike trains of neurons that generate rhythmic motor patterns. *Brain Research Bulletin* 21:529–538.

Nolfi, S., and D. Floreano. 2000. *Evolutionary Robotics: The Biology, Intelligence, and Technology of Self-Organizing Machines*. Cambridge, MA: MIT Press/Bradford Books.

North, G., and R. Greenspan, eds. 2007. *Invertebrate Neurobiology*. vol. 49. Cold Spring Harbor Monograph Series. Cold Spring Harbor, NY: Cold Spring Harbor Laboratory Press.

Parisi, D., and S. Nolfi. 1993. Neural network learning in an ecological and evolutionary context. In *Intelligent Perceptual Systems*, ed. V. Roberto, 20–40. Berlin: Springer.

Pfeifer, R., and J. Bongard. 2007. *How the Body Shapes the Way We Think: A New View of Intelligence*. Cambridge, MA: Bradford Books/MIT Press.

Philippides, A., P. Husbands, and M. O'Shea. 2000. Four dimensional neuronal signaling by nitric oxide: A computational analysis. *Journal of Neuroscience* 20 (3): 1199–1207.

Philippides, A., P. Husbands, T. Smith, and M. O'Shea. 2003. Structure based models of NO diffusion in the nervous system. In *Computational Neuroscience: A Comprehensive Approach*, ed. J. Feng, 97–130. London: Chapman and Hall/CRC Press.

Philippides, A., P. Husbands, T. Smith, and M. O'Shea. 2005. Flexible couplings: Diffusing neuromodulators and adaptive robotics. *Artificial Life* 11 (1–2): 139–160.

Philippides, A., S. Ott, P. Husbands, T. Lovick, and M. O'Shea. 2005. Modeling co-operative volume signaling in a plexus of nitric oxide synthase-expressing neurons. *Journal of Neuroscience* 25 (28): 6520–6532.

Pinaud, R., L. Tremere, and P. De Weerd, eds. 2006. *Plasticity in the Visual System: From Genes to Circuits*. New York: Springer.

Pitti, A., R. Niiyama, and Y. Kuniyoshi. 2010. Creating and modulating rhythms by controlling the physics of the body. *Autonomous Robots* 28 (3): 317–329.

Reil, T., and P. Husbands. 2002. Evolution of central pattern generators for bipedal walking in real-time physics environments. *IEEE Transactions on Evolutionary Computation* 6 (2): 10–21.

Santos, B., X. Barandiaran, and P. Husbands. 2012. Synchrony and phase relation dynamics underlying sensorimotor coordination. *Adaptive Behavior* 20 (5): 321–336.

Schmidhuber, J., D. Wierstra, M. Galiolo, and F. Gomez. 2007. Training recurrent networks by Evolino. *Neural Computation* 19 (3): 757–779.

Seth, A. 2005. Causal connectivity of evolved neural networks during behaviour. *Network: Computation in Neural Systems* 16 (1): 35–54.

Seth, A. K., A. B. Barrett, and L. Barnett. 2011. Causal density and integrated information as measures of conscious level. *Phil Trans R. Soc. A.* 369:3748–3767.

Seth, A. K., J. L. McKinstry, G. M. Edelman, and J. L. Krichmar. 2004. Visual binding through reentrant connectivity and dynamic synchronization in a brain-based device. *Cerebral Cortex* 14 (11): 1185–1199.

Shim, Y., and P. Husbands. 2010. Chaotic search of emergent locomotion patterns for a bodily coupled robotic system. In *Proceedings of ALife XI*, 757–764. Cambridge, MA: MIT Press.

Shim, Y., and P. Husbands. 2012. Chaotic exploration and learning of locomotion behaviours. *Neural Computation* 24 (8): 2185–2222.

Singer, W. 1999. Neuronal synchrony: A versatile code for the definition of relations? *Neuron* 24 (1): 49–65.

Smith, M., ed. 2007. *Stelarc:The Monograph*. Cambridge, MA: MIT Press.

Smith, T. M. C., P. Husbands, and M. O'Shea. 2003. Local evolvability of statistically neutral GasNet robot controllers. *Bio Systems* 69:223–243.

Smith, T. M. C., P. Husbands, A. Philippides, and M. O'Shea. 2002. Neuronal plasticity and temporal adaptivity: GasNet robot control networks. *Adaptive Behavior* 10 (3/4): 161–184.

Steingrube, S., M. Timme, F. Worgotter, and P. Manoonpong. 2010. Self-organized adaptation of a simple neural circuit enables complex robot behaviour. *Nature Physics* 6:224–230.

Suzuki, M., D. Floreano, and A. DiPaolo. 2005. The contribution of active body movement to visual development in evolutionary robots. *Neural Networks* 18 (5/6): 656–665.

Tanev, I., T. Ray, and A. Buller. 2005. Automated evolutionary design, robustness and adaptation of sidewinding locomotion of simulated snake-like robot. *IEEE Transactions on Robotics* 21:632–645.

Tononi, G., O. Sporns, and G. Edelman. 1992. Reentry and the problem of integrating multiple cortical areas: Simulation of dynamic integration in the visual system. *Cerebral Cortex* 2:310–335.

Tononi, G., O. Sporns, and G. Edelman. 1999. Measures of degeneracy and redundancy in biological networks. *Proceedings of the National Academy of Sciences of the United States of America* 96:3257.

von Twickel, A., A. Büschges, and F. Pasemann. 2011. Deriving neural network controllers from neuro-biological data: Implementation of a single-leg stick insect controller. *Biological Cybernetics* 104 (1–2): 95–119.

Varela, F., J. Lachaux, E. Rodriguez, and J. Martinerie. 2001. The brainweb: Phase synchronization and large-scale integration. *Nature Reviews. Neuroscience* 2 (4): 229–239.

Varela, F., E. Thompson, and E. Rosch. 1991. *The Embodied Mind.* Cambridge, MA: MIT Press.

Vaughan, E., E. Di Paolo, and I. Harvey. 2004 The evolution of control and adaptation in a 3D powered passive dynamic walker. In *Proceedings of the Ninth International Conference on the Simulation and Synthesis of Living Systems, Artificial Life IX*, ed. J. Pollack, M. Bedau, P. Husbands, T. Ikegami, and R. Watson, 139–145. Cambridge, MA: MIT Press.

Vargas, P., R. Moioli, L. Castro, J. Timmis, M. Neal, and F. Von Zuben. 2005. Artificial homeostatic system: A novel approach. In *Proceedings of the VIIIth European Conference on Artificial Life*, ed. M. Capcarrère, A. Freitas, P. Bentley, C. Johnson, and J. Timmis, 754–764. Berlin: Springer.

Vargas, P., R. Moioli, F. Von Zuben, and P. Husbands. 2009. Homeostasis and evolution together dealing with novelties and managing disruptions. *International Journal of Intelligent Computing and Cybernetics* 2 (3): 435–454.

Von der Malsburg, C. 1981. *The Correlation Theory of the Brain. Internal Report.* Gottingen, Germany: Max-Planck-Institute for Biophysical Chemistry.

Webb, B. 2001. Can robots make good models of biological behaviour? *Behavioral and Brain Sciences* 24 (6): 1033–1050.

Webb, B. 2009. Animals versus animats: Or why not model the real iguana? *Adaptive Behavior* 17 (4): 269–286.

Wheeler, M. 2005. *Reconstructing the Cognitive World.* Cambridge, MA: MIT Press.

Wiley, D., S. Strogatz, and M. Girvan. 2006. The size of the sync basin. *Chaos* 16:015103.

Wimsatt, W. 2002. Using false models to elaborate constraints on processes: Blending inheritance in organic and cultural evolution. *Philosophy of Science* 69 (3), Supplement: Proceedings of the 2000 Biennial Meeting of the Philosophy of Science Association. Part II: Symposia Papers: S12–S24.

Wood, J., and J. Garthwaite. 1994. Model of the diffusional spread of nitric oxide—Implications for neural nitric oxide signaling and its pharmacological properties. *Neuropharmacology* 33:1235–1244.

Zahedi, K., N. Ay, and R. Der. 2010. Higher coordination with less control—A result of information maximization in the sensorimotor loop. *Adaptive Behavior* 18 (3–4): 338–355.

3 Dynamical Analysis of Evolved Agents: A Primer

Randall D. Beer

3.1 Introduction

In many evolutionary robotics (ER) projects, the evolution of high-performing agents is just the beginning. Once successful agents have been evolved, we often want to *understand* how they work. For example, in my own work, in which evolutionary algorithms are used to explore the implications of a situated, embodied, and dynamical perspective on behavior and cognition, dynamical analysis of evolved model brain-body-environment systems using the mathematical tools of dynamical systems theory is central to the entire enterprise. Even in more engineering-oriented projects, it may be necessary to understand the range of conditions over which an evolved solution can be trusted to perform satisfactorily. In addition, the insights gained from analysis can sometimes be used to improve the performance of an evolutionary search (Mathayomchan and Beer 2002). Examples of the dynamical analysis of evolved agents include Beer 1995a; Husbands, Harvey, and Cliff 1995; Tani and Nolfi 1999; Beer 2003; Negrello and Pasemann 2008; Williams, Beer, and Gasser 2008; Izquierdo, Harvey, and Beer 2008; Izquierdo and Buhrmann 2008.

Analysis is not a monolithic activity; there is no standard mathematical procedure into which an evolved agent is dropped, some crank is turned, and out comes "understanding." Rather, analysis is a creative process, in which one actively engages the phenomenology of the evolved agents so as to answer particular questions with a given set of mathematical tools. The questions of interest can range from very specific inquiries about a particular property of a single evolved agent to very general examinations of the common features of large sets of successful agents. In addition, evolved agents can be analyzed at many different levels of description: the overall task, the behavior produced by the coupled brain-body-environment system, the agent-environment interactions that underlie the behavior of the coupled system, and the grounding of these interactions in specific neural, body, and environmental properties. Finally, many different mathematical tools can be employed, of which dynamical systems theory is only one. A good strategy is to begin an analysis with some specific question about a set of evolved agents and let this question guide the choice of level of analysis and mathematical tools, generalizing to a broader set of questions as understanding improves.

The goal of this chapter is to illustrate in a step-by-step manner the process of dynamical analysis of evolved agents. For this purpose, we study a walking agent consisting of a single leg in closed-loop interaction with a single neuron. Although the single-leg walking task is quite simple, it has also turned out to be incredibly rich. For example, we have used to it examine the conditions under which different pattern generator organizations evolve, the dynamical structure of these different organizations, the modular decomposition of central pattern generators, the interplay between neural and biomechanical properties in the generation of walking, and the impact of network architecture on performance and evolvability (Beer and Gallagher 1992; Beer 1995a, 1995b; Chiel, Beer, and Gallagher 1999; Beer, Chiel, and Gallagher 1999; Psujek, Ames, and Beer 2006). This combination of simplicity and richness makes it ideal for our pedagogical purposes here. In this chapter, we consider how evolved walking agents work and what this tells us about where in parameter space they might be found. The analysis described here is fairly qualitative in nature; a much more rigorous and extensive analysis of this system with full mathematical details can be found in Beer 2010.

3.2   Model

The single-leg model that we employ is illustrated in figure 3.1 (Beer and Gallagher 1992). A leg is composed of a segment of length L connected to the body by a joint actuated by two opposing "muscles" (or effectors) BS and FS, for controlling the back and forward swing of the leg respectively, and a binary foot FT. When the foot is "up" *(swing phase)*, any torque produced by the muscles serves to swing the leg along an arc relative to the body, with a maximum angular acceleration of α_{max} and angular limits of $[\phi_{min}, \phi_{max}]$. When a swinging leg reaches this angular limit, it comes to an abrupt stop. When the foot is "down" *(stance phase)*, any torque produced by the muscles applies a translational force to the body under Newtonian mechanics, with a maximum acceleration of a_{max}. Note that during stance phase the leg stretches between the body joint and the stationary foot as the body moves; the horizontal distance between the joint and the foot is labeled d. The leg is only able to generate force over a limited angular range of motion of $[\phi_{min}, \phi_{max}]$ (modeling how mechanical advantage changes with limb geometry). A stancing leg exceeding these limits can still provide support, but only within vertical limits of $[x_{min}, x_{max}]$ (modeling skeletal constraints). When a stancing leg reaches these hard kinematic limits, forward motion comes to an abrupt stop (modeling a loss of postural stability). Note that, when a stretched stancing leg lifts its foot, the leg immediately snaps back to the swing angular limits of $[\phi_{min}, \phi_{max}]$

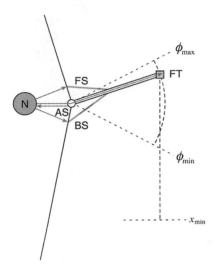

Figure 3.1
Configuration of the model agent. The single neuron N sends output to the forward swing (FS), backward swing (BS), and foot (FT) effectors, and in turn receives input from the angle sensor (AS). The dashed gray lines mark the path of the foot during swing (curved path) and stance (straight path), assuming that the foot is put down at the leg angle shown. The various angle limits mentioned in the main text are shown as black dashed lines.

(modeling the passive restoring force of muscle). Each leg also possesses an angle sensor AS whose output is proportional to the angular deviation of the leg from perpendicularity to the long axis of the body. The leg parameter values utilized in this work were $L = 15$, $\alpha_{max} = 1/40$, $a_{max} = 1/20$, $\phi_{max}, \phi_{min} = \pm\pi/6$, and $x_{max}, x_{min} = \pm20$.

The model body was coupled to a continuous-time recurrent neural network (CTRNN). The operation of a standard N node CTRNN is described by equations 3.1 and 3.2.

$$\tau_i \frac{dy_i}{dt} = -y_i + \sum_{j=1}^{N} w_{ji} \sigma(y_j + \theta_j) + S(t) \tag{3.1}$$

$$\sigma(x) = 1/(1 + e^{-x}) \tag{3.2}$$

where y_i is the mean membrane potential of the ith neuron, τ_i is the neuron's membrane time constant, w_{ji} is the strength of the synaptic connection from the jth to the ith neuron, θ_j is a bias term, $\sigma(x)$ represents the neuron's mean firing rate, and $S(t)$ is a sensory input signal.

In the work described in this chapter, we have only a single CTRNN neuron that receives as input the sensory signal $S = S_w \phi$ from the angle sensor, where $S_w = 30/\pi$.

The output $o = \sigma(y + \theta)$ of this neuron drives the leg muscles and foot. Specifically, FT is "up" when $o \le 1/2$ and "down" when $o > 1/2$, and the neuron output scales α_{max} during swing and a_{max} during stance. If we assume that the leg parameters are all fixed to constant values as described, then the model dynamics depends on only three neural parameters: the self-weight w, the bias θ, and the time constant τ of the single CTRNN neuron.

If we work in the output space of the neuron and the angular coordinates of the leg joint, then the complete model can be succinctly defined by expressing the swing and stance dynamics of the walking agent as separate sets of differential equations that are switched between whenever the foot changes state. The relevant state variables are the neuron output o, the leg angle ϕ at the joint, and the leg angular velocity ω.

The fitness measure we employ to characterize the walking performance of a legged agent is the average forward velocity of the body. We have typically computed this average velocity in two ways. During evolution, *truncated fitness* is computed by integrating the model for a fixed length of time and then dividing the total forward distance covered by the evaluation duration. During analysis, *asymptotic fitness* is computed by integrating the model for a fixed length of time to skip transients and then calculating its average velocity for one stepping period (with a fitness of 0 assigned to nonoscillatory circuits). Since the focus of this chapter is on analysis, we will employ asymptotic fitness throughout.

3.3   The Operation of a Typical Agent

An excellent place to begin a dynamical analysis is with a detailed examination of the operation of a single highly fit individual. A detailed understanding of one agent provides a strong foundation for asking more general questions. For this purpose, we focus in this section on the best individual found out of 250 evolutionary searches. The evolved parameter values for this agent were $w = 16$, $\theta = -4.4876$, and $\tau = 1.3956$.

When analyzing an evolved agent, I prefer to begin with a visualization of its dynamics. Once some intuition for what is going on has been developed, it is often straightforward to translate this understanding into equations that can then be utilized for subsequent analysis. For our purposes here, we need a visualization that illustrates the dynamics of the leg, the dynamics of the neuron, and the effect that each of these dynamics has on the other. Figure 3.2a illustrates the operation of the best evolved walker in (ϕ, o) space (since this system is Newtonian, the third state variable ω is just the time derivative of ϕ).

The black curve shows the (ϕ, o) trajectory of the system over one step cycle. Stance phase begins at the right-hand side of this plot, when the output of the neuron exceeds $1/2$ and the foot is put down. The trajectory then moves to the left as the leg angle decreases during stance phase. Stance phase ends when the output of the neuron falls

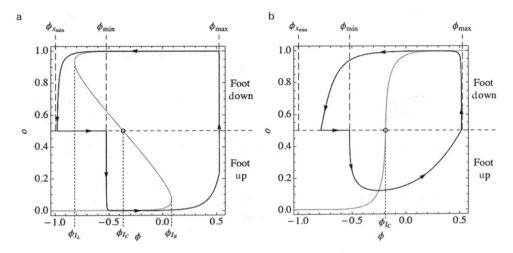

Figure 3.2
Operation of two evolved walkers. (A) Operation of the best evolved agent, with parameters $w = 16$, $\theta = -4.4876$, $\tau = 1.3956$. (B) Operation of the best-evolved agent with $w < 4$. The parameters of this agent are $w = 3.6224$, $\theta = 0.0051$, $\tau = 8.4555$. In parts a and b, black curves denote stable limit cycles, small circles indicate unstable equilibrium points, and a neuron's steady-state input/output curve is shown in gray.

below 1/2 and the foot is lifted. Since the leg has stretched beyond ϕ_{min} during stance, it snaps back to this angle when the foot goes up. The leg angle then increases during the swing phase until the neuron output once again exceeds 1/2, at which point the foot goes back down and the cycle repeats.

What role does the neuron play in this walking cycle? In order to answer this question, we need to visualize how the neuron's output depends on the sensory input $S_w \phi$ that it receives from the leg. The S-shaped gray curve in figure 3.2a shows the neuron's steady-state input/output (SSIO) curve. This curve gives the locations of the equilibrium points of the neuron as a function of its sensory input. Note that the SSIO curve is folded, indicating a region of bistability, with the upper and lower branches of the SSIO corresponding to stable equilibrium points and the middle branch corresponding to an unstable equilibrium point. When the leg angle is large enough that the sensory input exceeds the right edge of the fold, only the upper branch of the SSIO exists and the neuron state is attracted to that equilibrium point. Likewise, when the leg angle is sufficiently negative, only the lower attractor exists and the neuron state is attracted to it. When the leg angle is such that the sensory input falls within the fold, which stable branch of the SSIO the neuron state will be attracted to depends on its initial state. If the state begins above the middle unstable branch, then it will be attracted to the upper stable branch. If it begins below the middle branch, then the state will be attracted to the lower branch.

Using the diagram in figure 3.2a, we can understand the basic operation of the best evolved walker as follows. At the beginning of stance phase, the leg angle is such that only the upper equilibrium point of the neuron exists, attracting the neuron output toward 1. Once $o > 1/2$, the foot goes down and a stance phase begins, causing ϕ to decrease. As the leg angle approaches $\phi_{x_{min}}$, the sensory input to the neuron passes below the left edge of the SSIO fold and only the lower equilibrium point of the neuron is stable, attracting the neuron output toward 0. Once $o \leq 1/2$, the foot lifts and snaps back to ϕ_{min} and then o continues on toward 0, initiating a swing phase that eventually increases the leg angle past the point where the sensory input exceeds the right edge of the SSIO fold, leading to another stance phase. Thus, the stepping cycle arises from a reciprocal interaction between the shape of the neuron's SSIO curve and the movement of the leg, with any given neuron output serving to drive the leg toward an angle that will eventually cause the neuron output to switch.

From previous analysis of CTRNNs (Beer 1995b), we know that a folded SSIO only occurs when $w > 4$. Is it possible to have a functioning walking agent when $w < 4$? Examining the results of a number of evolutionary searches, we find that functional walking agents with $w < 4$ do in fact occur, although their fitness tends to be lower than agents with $w > 4$. Figure 3.2b illustrates the operation of the best such agent found among 250 evolutionary searches. Note that its basic operation is very similar to the agent shown in figure 3.2a. The only real difference is that, because its SSIO curve is not folded, the leg angle only has to get to a point where the sensory input is to the right of the center point of the SSIO in order to make a transition from swing to stance. Similarly, the leg angle must get to a point where the sensory input is to the left of the center point in order to make the transition from stance to swing.

3.4   The Location of Functional Walkers in Parameter Space

With the intuition that we developed in section 3.3, we are now in a position to ask more general questions about this model agent. The question we will focus on in the remainder of this chapter is: Where in (w, θ, τ) parameter space can functional walking agents be found? By "functional," I mean agents that are capable of rhythmic stepping. Note that this question represents an enormous leap in scope, from the operation of individual walking agents to the structure of the space of all possible instances of this model agent.

Our first step in answering this question is to examine the ways that the interaction between the leg and the neuron's SSIO can fail to produce rhythmic stepping. To do this, we need to recall how a CTRNN's SSIO depends on its parameters (Beer 1995b). First, the shape of the SSIO curve is independent of τ, so we will temporarily ignore this parameter. Second, increasing (decreasing) θ shifts the SSIO curve left (right) in input space. Third, as we have already seen, w determines the degree of folding of the SSIO curve. It also influences the location of the fold in input space, with larger w values moving the fold to the left.

One way in which the walking agent shown in figure 3.2a can fail to produce rhythmic stepping is if the sensory input from the leg at the most forward angle ϕ_{max} the leg can reach during swing is insufficient to clear the right edge of the SSIO fold. In this case, the leg will become "stuck" in a permanent swing phase, with the neuron output unable to make the transition toward 1 and initiate a stance phase. Since decreasing θ shifts the SSIO curve to the right, this constraint determines a w-dependent lower bound on θ that must be satisfied in order for successful stepping to occur. We will call this boundary B1.

By similar reasoning, we can derive four other constraints on functional walkers, each of which gives rise to another boundary in parameter space. For example, in order for the agent shown in figure 3.2a to make the transition from stance to swing, the left edge of the SSIO fold must be greater than the most negative angle $\phi_{x_{min}}$ that the leg can reach during stance. We will call the boundary defined by this constraint B2. Boundary B3 is given by the constraint that the angle of the leg after snapping back to ϕ_{min} when the foot is lifted must lie to the left of the center of the SSIO curve in order for the neuron output to complete its transition to the lower stable branch of the SSIO. When $w < 4$, an examination of figure 3.2b gives rise to two more boundaries. Boundary B4 is given by the constraint that the sensory input from the maximum leg angle ϕ_{max} reached during swing must clear the center of the SSIO curve in order for the swing-to-stance transition to occur. Boundary B5 is given by the constraint that the sensory input from the minimum leg angle $\phi_{x_{min}}$ reached during stance must clear the SSIO center in order for the stance-to-swing transition to occur.

The next step in our analysis is to derive equations for the five boundaries determined by the constraints that we have outlined. We will illustrate this process with boundary B1. The constraint underlying B1 can be written as $\phi_{I_R} < \phi_{max}$. The critical boundary thus occurs when $\phi_{I_R} = \phi_{max}$. We know from previous analysis of CTRNNs that the right edge of the SSIO fold in input space is given by the expression $I_R(w) - \theta$ (Beer 1995b). Since the input to the neuron from the leg is given by $S_W \phi$, the leg angle ϕ_{I_R} corresponding to the right edge of the fold is given by $S_W \phi_{I_R} = I_R(w) - \theta$ or $\phi_{I_R} = (I_R(w) - \theta)/S_W$. Thus, B1 satisfies the equation $(I_R(w) - \theta)/S_W = \phi_{max}$ or $\theta = I_R(w) - S_W \phi_{max}$. Similar reasoning can be used to derive equations for the other four boundaries, although there are some subtleties with B2 that we will not go into here (Beer 2010).

It is now time to test our analysis against the actual structure of the fitness space for this model agent and to determine whether we have missed any important

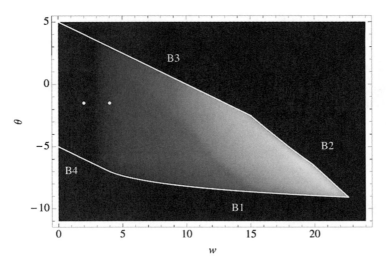

Figure 3.3
A fitness slice through (w, θ) parameter space at $\tau = 0.5$. The density plot in the background indicates the fitness of each parameter combination, with brighter areas corresponding to higher fitness. The white curves superimposed on the density plot indicate the theoretical boundaries described in the main text. Note that no theoretical boundary corresponds to the left edge of the density plot. The two white points give the parameter space locations of the two walkers shown in figure 3.4.

features. Toward this end, figure 3.3 shows our predicted boundaries as white curves superimposed over a density plot of fitness space for the (w,θ) slice at $\tau = 0.5$. As it turns out, boundary B5 is subsumed by the tighter boundary B3, so only B1–B4 are shown. Note that our theoretical boundaries provide an excellent fit to the lower, upper, and right edges of the high-fitness region in this slice.

However, something is definitely wrong on the left side of figure 3.3. Although our analysis predicts no boundary there, the fitness density plot clearly exhibits a sharp left edge. Since our preceding analysis has accounted for all the possible ways that the neuron's SSIO can interact with the various angle limits of the body to prevent rhythmic stepping, something else must be going on to create this left boundary that we have not yet considered.

In order to determine what we missed, we will return to looking at particular examples in order to build intuition. Figure 3.4 shows the operation of two walkers that lie just to the left and just to the right of this new boundary (their locations in parameter space are marked by white dots in figure 3.3). In the walker to the right of the new boundary (figure 3.4b), we see a stable limit cycle surrounding an unstable equilibrium point just as we did in both walkers shown in figure 3.2. Note that the

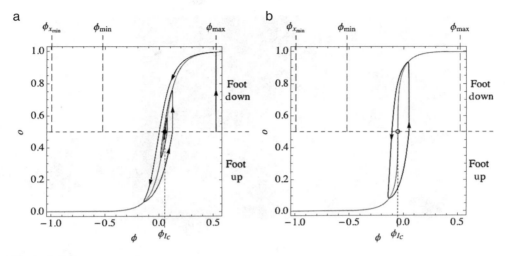

Figure 3.4
Operation of two walkers whose parameter space locations are indicated by the left (a) and right
(b) white points in figure 3.3. The labeling conventions are the same here as in figure 3.2, except
that the black dot in part a corresponds to a stable equilibrium point. As w increases from part
a to part b, the system undergoes a Hopf-like bifurcation, in which a stable spiral equilibrium
point loses stability, giving rise to a stable limit cycle.

limit cycle is rather narrow, corresponding to steps that cover only a small angular
range. In contrast, in the walker to the left of the new boundary (figure 3.4a) we see
only a single stable equilibrium point. In this case, transient rhythmic stepping occurs,
but it eventually decays; the agent only takes a limited number of increasingly smaller
steps until coming to a standstill.

What happens between these two walkers to create the left boundary? The change
between figure 3.4a and figure 3.4b is reminiscent of a Hopf bifurcation, in which a
stable spiral equilibrium point loses stability and gives off a stable limit cycle in
the process (Strogatz 1994). We will use the term "Hopf-like" to describe this bifurcation because, due to the switch-like character of the foot, the technical conditions
of a Hopf bifurcation are not satisfied in this system (di Bernardo et al. 2008).
Nevertheless, although we will not go into the details here, it is possible to write an
equation for the curve in parameter space along which this bifurcation occurs and to
solve this equation numerically (Beer 2010). The resulting boundary, labeled B6, is
shown in figure 3.5. Note that this curve matches the left edge of the density plot
perfectly. Thus, at least for the $\tau = 0.5$ slice that we have considered here, we have
developed a complete understanding of the location and layout of functional walkers
in parameter space.

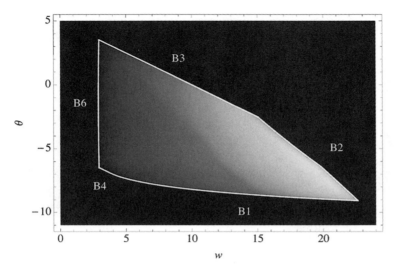

Figure 3.5
Adding the theoretical curve of Hopf-like bifurcations B6 to figure 3.3 completely accounts for the shape and location of the region of functional walkers in the parameter space for $\tau = 0.5$.

3.5 Discussion

This chapter has briefly illustrated the process by which one goes about dynamically analyzing evolved agents. We began with the very specific question of how the best evolved agent worked. This necessitated finding a way to visualize the walking dynamics that made explicit the interaction between key neural and mechanical properties of the agent (figure 3.2a). The study of this agent led us to consider a second agent for which a key neural property differed (figure 3.2b). The intuition gained from the examination of these two agents then allowed us to hypothesize some general constraints on functional walking agents, which were then made mathematically precise. Testing these predictions against a fitness slice through parameter space, we found that, although our predictions were generally quite accurate, there was one major discrepancy (figure 3.3). This required us to return to the examination of specific agents in order to understand the nature of the discrepancy (figure 3.4). With the insight gained from these additional examples, we were then able to state an additional constraint and thus fully account for the location and shape of the region of functional walkers in parameter space (figure 3.5).

If the flow of this chapter has a bit of the feel of solving a mystery, then it will have succeeded in communicating something of the nature of analysis. A "crime" has been committed (some sets of neural parameters lead to rhythmic stepping and some do not). We collect clues (e.g., how a particular successful walker operates) and interrogate suspects (e.g., what happens if this parameter is changed?), gathering evidence for making the case that our explanation of what really happened is correct. Even though the crime, clues, suspects, and evidence may differ significantly from one set of evolved agents to the next, the overall logic by which an analysis proceeds has many similarities from case to case.

In the interests of accessibility, this chapter has downplayed the mathematical details underlying our analysis. However, it is important to recognize that these details are crucial to the entire endeavor. The mathematics is what allows us to transform intuitive understanding into precise quantitative descriptions. It is what makes it possible to move from simulations of specific agents to a rigorous general theoretical framework for understanding the structure of the space of all possible instances of a given model agent. Despite the fact that the mathematics can be a high barrier for some, it provides essential tools for disentangling the complex causal mechanisms operating in evolved brain-body-environment systems.

Regarding the specific walking agent considered in this chapter, there is much more that can be done (Beer 2010). For example, it is possible to characterize the dependence of the boundaries we have analyzed in this chapter on τ. Although boundaries B1, B3, and B4 are independent of τ, boundaries B2 and B6 vary in interesting ways with the neuron time constant. In addition, it turns out that the left edge of the density plot is defined not only by the curve B6 of Hopf-like bifurcations, but also another bifurcation that comes into play at larger τ values. As can be seen in figure 3.5, there is also interesting internal structure to the high-fitness region that can be characterized. Finally, a similar analysis could be applied to walking agents with higher-dimensional parameter spaces, including central pattern generators (in which an intrinsically oscillatory neural circuit drives walking) and mixed pattern generators (in which a walking pattern arises from the interaction between a neural oscillator and rhythmic sensory feedback from the leg).

References

Beer, R. D. 1995a. A dynamical systems perspective on agent-environment interaction. *Artificial Intelligence* 72:173–215.

Beer, R. D. 1995b. On the dynamics of small continuous-time recurrent neural networks. *Adaptive Behavior* 3 (4): 471–511.

Beer, R. D. 2003. The dynamics of active categorical perception in an evolved model agent. *Adaptive Behavior* 11:209–243.

Beer, R. D. 2010. Fitness space structure of a neuromechanical system. *Adaptive Behavior* 18 (2): 93–115.

Beer, R. D., H. J. Chiel, and J. C. Gallagher. 1999. Evolution and analysis of model CPGs for walking II. General principles and individual variability. *Journal of Computational Neuroscience* 7:119–147.

Beer, R. D., and J. C. Gallagher. 1992. Evolving dynamical neural networks for adaptive behavior. *Adaptive Behavior* 1:91–122.

Chiel, H. J., R. D. Beer, and J. C. Gallagher. 1999. Evolution and analysis of model CPGs for walking I. Dynamical modules. *Journal of Computational Neuroscience* 7:99–118.

di Bernardo, M., C. J. Budd, A. R. Champneys, and P. Kowalczyk. 2008. *Piecewise-Smooth Dynamical Systems: Theory and Applications*. Berlin: Springer.

Husbands, P., I. Harvey, and D. Cliff. 1995. Circle in the round: State space attractors for evolved sighted robots. *Robotics and Autonomous Systems* 15:83–106.

Izquierdo, E., and T. Buhrmann. 2008. Analysis of a dynamical recurrent neural network evolved for two qualitatively different tasks: Walking and chemotaxis. In *Proceedings of the 11th International Conference on Artificial Life*, ed. S. Bullock, J. Noble, R. A. Watson, and M. A. Bedau, 257–264. Cambridge, MA: MIT Press.

Izquierdo, E., I. Harvey, and R. D. Beer. 2008. Associative learning on a continuum in evolved dynamical neural networks. *Adaptive Behavior* 16:361–384.

Mathayomchan, B., and R. D. Beer. 2002. Center-crossing recurrent neural networks for the evolution of rhythmic behavior. *Neural Computation* 14:2043–2051.

Negrello, M., and F. Pasemann. 2008. Attractor landscapes and active tracking: The neurodynamics of embodied tracking. *Adaptive Behavior* 16:196–216.

Psujek, S., J. Ames, and R. D. Beer. 2006. Connection and coordination: The interplay between architecture and dynamics in evolved model pattern generators. *Neural Computation* 18:729–747.

Strogatz, S. H. 1994. *Nonlinear Dynamics and Chaos*. New York: Addison-Wesley.

Tani, J., and S. Nolfi. 1999. Learning to perceive the world as articulated: An approach for hierarchical learning in sensory-motor systems. *Neural Networks* 12:1131–1141.

Williams, P. L., R. D. Beer, and M. Gasser. 2008. An embodied dynamical approach to relational categorization. In *Proceedings of the 30th Annual Conference of the Cognitive Science Society*, ed. B. C. Love, K. McRae, and V. M. Sloutsky, 223–228. Washington, DC: Cognitive Science Society.

4 Evolutionary Pathways

Inman Harvey and Ezequiel A. Di Paolo

4.1 Introduction

Heredity, in both natural and artificial evolution, is often pictured in the form of a tree. This may be at the level of individuals, where the lines trace pathways back through parents and grandparents to distant ancestors. It could be at a higher level, where the ancestral relationships traced are those of the species, genera, and orders. This simple and compelling picture is made somewhat muddier by a belated recognition that there may be interesting amounts of horizontal gene transmission between many of these branches. Nevertheless, the concept of the evolutionary pathway may be useful for many purposes. For people using artificial evolution, as in evolutionary robotics (ER), one question that may be asked is this: given the present situation of some evolving population, and given the desired end goal, what pathways are there that evolution may take and is there any way that we can persuade evolution to head along the faster ones?

That question, in its very general form, is what we are tackling in this chapter. In doing so we shall move back and forth between rather abstract discussion of evolutionary search spaces and practical tips and hints that may be useful for people designing and applying evolutionary algorithms for some specific problem. Sometimes such tips have been found useful by practitioners and are passed on by word of mouth, yet do not find their way into the published literature. As heuristics, they are sometimes crude and unproven. This chapter is not intended as an exhaustive classification of evolutionary pathways; it should rather be considered as an album of postcard views drawn from several ER travelogues, in the hope that some of them may inspire and influence future explorers.

We discuss the various features of evolutionary paths that can make them more or less easy for evolution to navigate. These are illustrated by some case studies, and where the lessons or intuitions thus suggested may have some wider applicability they are summarized in the form of "Travel Tips."

4.2   Where to Start From

The apocryphal Irish peasant, accosted in a rural lane by an American tourist in a rental car asking for the best route to Dublin airport, scratched his head and replied: "If I was going there, I wouldn't be starting from here." There are often good choices and bad choices for a starting population. Initializing the population at random sounds as though it should be simple, but there are significant subtleties. The Bayesian rationale behind initializing a population at random is that if we have no prior knowledge of where the solution or solutions may be found, then we should not bias the search in any way—but where we do have some prior knowledge we may be justified in incorporating it.

In evolutionary robotics it is common for all or part of the genotypes to be specifying real values for things like weights and biases of a neural network, time parameters, sizes of body parts, and so on. When any of these values are potentially unbounded in one or both directions, the search space is infinite in size. We must scatter the initial population over some bounded region that includes where the goal will be. A careless guess for the size of these bounds could be orders of magnitude too large or too small. The programmer may be hoping for evolution to do most of the work, but a reality check—needing some ability to reason with orders of magnitude—could help stop the search in the right ballpark. In general, if some of the parameters tend to evolve toward one of the extremes of their range, this may be a sign that the parameter range should be changed accordingly. Incidentally, when a range is indeed given hard boundaries then a decision has to be taken as to how to treat mutations that mutate values "out of range"; one solution often found effective is to "reflect" such a mutation back off the end-of-range value as if it was a reflecting wall.

In the absence of better knowledge, parameter ranges should be searched uniformly without bias. This is often done using linear mapping of genotype values onto the parameter range. However, the meaning of an unbiased search can depend on the parameter. When evolving CTRNNs (Beer 1995a, 1995b), time parameters τ are significant, and should often be allowed to range over several orders of magnitude, for instance from milliseconds to hours as the performance of the task may involve a variety of timescales (up to and including the timescale of the whole performance, which can include several trials when, for instance, evolving learning behavior). One trick to facilitate this is to encode time parameters logarithmically on the genotype. A value t on the genotype gets translated to τ via a formula such as $\tau = e^t$, or $\tau = 10^t$. This allows t to range through positive and negative values, while τ is always positive. It means that an initial population, drawn from a flat distribution in t values, will be realistically scattered across several orders of magnitude in τ values, and subsequent mutations also behave more appropriately.

4.2.1 CTRNNs or CTSNs

The choice of initial bounds, and the use of logarithmic encoding where appropriate, are examples where we do have some prior knowledge and are justified in incorporating this. A further example of this is the use of center-crossing (Beer 1995b; Mathayomchan and Beer 2002), where an understanding of the dynamics of CTRNNs is used to inform the choice of an initial population. The mathematical formalism of a CTRNN covers dynamical systems in general, and the inclusion of NN for neural networks in the name has misled some people who have been puzzled that the "weights" involved are fixed. The description CTSN, for continuous time sigmoidal networks (Beer and Daniels 2010), may have less misleading assumptions attached to it. Whichever term is used, the "nodes" of such networks refer to any modeled variable of a dynamical system—and could, for instance, include the weights of a conventional NN. The benefits of using CTRNNs or CTSNs as a default architecture for an evolved robot "brain" include:

1. They provide universal approximators for any smooth dynamical system (Funahashi and Nakamura 1993).
2. They are well understood, with a lot of accumulated experience and some analytical results.
3. They are easy to implement.
4. They are relatively easy to analyze.
5. They give good building blocks for dynamical systems, e.g., their activations are squashed so that values do not shoot off to infinity.

Each neuron in a CTRNN has a nonlinear activation function, with a central region (defined by the value of the bias term) that specifies where its output is most sensitive to varying input values. For summed input values significantly higher or lower than this, the nonlinear activation function saturates close to its upper or lower bounds, and ceases to be sensitive to further variations. In effect it is then switched (almost) fully on or fully off and plays (almost) no further role in the dynamics of the network as a whole. Small variations in connection weights associated with such a saturated neuron would make negligible changes to its function, and hence it becomes difficult for evolution to "explore viable pathways." Beer (1995b) conjectured that if an initial population was manipulated so as to ensure that all the neurons started off in their sensitive regions, then this would maximize the prospects for evolution. This can be done by ensuring that there is an equilibrium point for the network with every neuron's activation at its exact center of symmetry, in other words, with an output of 0.5 halfway between the saturation levels of 0.0 and 1.0.

A simple way to achieve this in an initial population that is otherwise random is to initialize the bias term for each neuron, as a function of the connection weights, so as to obey the "center-crossing" condition (Beer 1995b). In other words, although the connection weights are initialized randomly, the biases are initially constrained to meet this condition. Thereafter evolution is free to change (through mutation) all the parameters including these biases. "The richest dynamics should be found in the neighborhood of the center-crossing networks in parameter space, and one would expect that an evolutionary algorithm would benefit from focusing its search there" (Beer 1985b). In addition, if it were convenient for some neurons to work in the saturated regions, evolution could easily find suitable parameters to achieve this from such a starting point, while the opposite move from saturated to more dynamical regions is trickier.

In (Mathayomchan and Beer 2002) this strategy is compared against pure random initialization in evolving central pattern generators for a simulated walking creature, and found to give significantly superior performance. Analysis of CTRNNs (Beer and Daniels 2010) also shows that under some circumstances new regimes of dynamics become available when self-weights are allowed to exceed a value of 4 allowing neurons to become bistable, hence restrictions of weights to smaller values than this can limit the richness of possibilities. Randall Beer provides a primer (chapter 3, this volume) with pointers to such studies. It pays to pay attention to where an evolutionary pathway starts.

Travel Tip (TT) 1: Choose limits on initial values in a random population so as to span the full range in which you expect to find good results.

TT2: Consider "reflecting" mutations that take a value out of bounds as if the bounds were a reflecting wall.

TT3: Where an evolved value may scale several orders of magnitude, consider scaling it logarithmically on the genotype.

TT4: Where possible, ensure the initial population is free to evolve in any direction; for example, for CTRNNs, use the center-crossing method.

4.3   Where We Are Heading

In biology, we are often interested in pathways leading to the here and now of some particular individual or species. In evolutionary robotics we are typically looking to the future, and must decide what would count as a satisfactory end goal. This goal may be more or less explicit; it may demand solutions to specific tasks or the satisfaction of broad viability constraints (explore the world, survive, and so on). If it is some uniquely specified robot behavior, there are probably very many different genetic specifications, different genotypes, that could generate this. Usually we are seeking some satisfactory class of behaviors, rather than something unique, so the target is wider still. In terms of evolutionary search, we are seeking the uplands of some fitness landscape rather than one unique point at the peak of Mount Everest.

Failure to realize this may underlie the common, but false, intuition that assumes big search spaces must be worse than small search spaces. After all, surely searching for a single needle in a large haystack is worse than searching for a single needle in a small haystack? But actually, if the density of needles is the same in each case, that intuition is misguided. Sometimes it may be the big haystack that is more navigable than the small one. Choices of genetic encoding will translate through to the size of the evolutionary search space. Any attempts to make the search space small, while ignoring consideration of the density of acceptable solutions or the navigability of the fitness landscape, are likely to be counterproductive.

If a fitness landscape only has two levels of fitness—in haystack terms, needles equals good and hay equals bad—then no search can be better than random search. The expected time to success depends only on the ratio of needles to hay. For evolutionary search to have some advantages, fitness should be measured in shades of gray and not just black and white. If one relates these different shades of fitness to contour lines on the fitness landscape, then any evolutionary pathway up the hills corresponds to passing through intermediate subgoals, which should be easier to find, en route to the more elusive uplands. Sometimes we can reshape the landscape by introducing intermediate targets, or stepping-stones, and we give some examples below. An intermediate target that is wider than the ultimate bull's-eye will of course be easier; and as archers know, surrounding the bull's-eye with concentric circles is one more way to add shades of gray to the fitness function.

TT5: Big search spaces may be as easy as, or sometimes even easier than, smaller search spaces—it is the density of satisfactory solutions that should be considered.

TT6: Fitness functions in shades of gray are usually easier than discrete black-and-white values.

4.4   Incremental Evolution

A classic strategy for tackling a difficult problem is to first carve it up into smaller ones: divide and conquer. When designing robots, such division is typically at the level of mechanism. Development of a motor mechanism can go ahead largely independently of any development of a vision system, since although each such module may be complex internally we expect the interactions between them to be rather simpler and easier to understand. But though humans find it easier to design complex systems in such a modular fashion, it is not so clear whether evolved systems, naturally or artificially evolved, have the same overwhelming need for modularity. It is significant that evolutionary robotics can and does produce evolved systems whose complexity of interactions is too difficult for humans to analyze and understand.

But there is a different way to carve up a large problem into smaller ones. A complex behavior can be seen as composed of simpler behaviors. Insofar as we can see, with hindsight, the evolutionary pathways from the origin of life to present-day mammals, we see phylogenetic development from relatively small and simple organisms with relatively simple behaviors through successive increases in complexity and variety of mechanisms and of behaviors. Necessarily every single individual in any one of these pathways must have been viable in its own right. This thought inspired Brooks (1991a, 1991b) to propose a subsumption architecture as a method for the incremental design of robots.

The way in which an evolutionary algorithm can be applied to such an incremental process and the use of such methods in ER were proposed in Harvey (1992), Harvey, Husbands, and Cliff (1993), and a body of work following from this. These resulted in the SAGA principles (Harvey 2001) of balancing mutation rates against selection pressure so as to optimize the exploitation/exploration balance in such scenarios. Incremental evolution is now quite widely practiced. Sometimes it is called robot shaping (Dorigo and Colombetti 1998), or incremental shaping (Auerbach and Bongard 2011). The term "scaffolding" is used in developmental psychology to refer to the framework that guides a child's educational development, and can naturally extend to ER. It is clear that the choice of scaffolding methods can affect the ease or difficulty of evolutionary progress (Auerbach and Bongard 2011).

There are two separate issues to worry about when doing incremental evolution. First, one must assign the appropriate stepping-stones leading in the right direction and a suitable distance apart. This translates to designing a new fitness function for the next stage, to bias selection in favor of the desired behavior. An example can be seen in Vickerstaff and Di Paolo (2005), where an evolved agent was required to navigate via a series of beacons, with successive stages being also incremental in terms of complexity—evolving simple phototaxis first, then phototaxis to multiple sources, and finally homing to a nest.

The second issue is to decide whether new mechanisms are appropriate, for instance sensors, or extra brainpower in the form of a larger neural network, and if so, how to handle grafting such new material on to what is inherited from the previous stages.

Vaughan, Di Paolo, and Harvey (chapter 11, this volume) present a detailed case study on how incremental evolution was used to design a bipedal walking robot. An important lesson from this was that the effectiveness of results at one incremental stage, and the problems encountered there, informed the decision of what the next stepping-stone should be and what sorts of extra facilities might be needed to achieve that next step. So the evolutionary pathway was not prespecified from the beginning, but instead was laid out as evolution progressed in a dialogue between the analysis of the engineer and the high-dimensional parameter search of the genetic algorithm (GA).

Vaughan (2007) used a significantly useful trick in this and related work, which only received a passing mention in his thesis. At a new stage in the incremental evolution, if there was for example a new sensor with an associated neural network to be added to the previously evolved system, then the weights and biases of this new neural network could be set at random. But the trick was to make sure all the connection weights between this new neural network and the previously evolved one were set to zero. In consequence, the new sensor and its locally associated neural network had absolutely no effect on the previously evolved system. Nevertheless further evolution was allowed to modify thereafter the parameters for the old network, for the new network, and for their interconnections.

By doing it this way, the potential new functionality was added in an initially neutral manner. As with neutral networks in general (to be discussed later in this chapter), the extra dimensionality opened up possibilities for new pathways. If the engineer's analysis and intuitions in making available the new sensor and deciding where it should connect up to the existing network were not sound, if incorporation of these new components would actually lead to loss of fitness, then evolution "had the option" of keeping the interconnection weights at zero and the new part effectively was ignored. But if the intuitions were useful, then the new component part could coevolve together with the previous system and interconnections between them.

TT7: Consider intermediate targets, less specialized than the ultimate goal, as stepping-stones.

TT8: When adding new component parts to the robot or its "brain" at a new stepping-stone, consider having them initially connected to the preexisting system in a neutral way (e.g., with zero weights), so as to preserve the previous functionality.

4.5   Showing the Way

Setting out a pathway of stepping-stones, through incremental changes in the tasks on which evolving robots are evaluated, is one way that the engineer can influence the direction taken. But there is another way of implicitly encouraging evolution to head toward the next goal via some particular direction. In this next example, the engineers crafted their fitness function to make a crucial factor salient. This could be related to the biological concept of exaptation, where something that "will be useful in the future," such as feathers on a flying creature, may have been originated with some different role, such as providing warmth. In natural evolution the Blind Watchmaker has no foresight, so that when one feature is exapted to take on a new role, this is serendipitous. But in artificial evolution, where there is scope for foresight, the Watchmakers may be partially sighted.

In one example (Tuci, Quinn, and Harvey 2002) experimenters introduced an artificial bias in the evolutionary process that was stimulated by a lack of success in evolving a second-order behavior. The task for a (simulated) robot was to navigate in a 2D
arena toward a distant target, placed on the floor at random at either the left or right
end of the arena. The target was not visible until reached, and the only visible indication of which end to head for was a light signal. Over some sequences of trials the
light was associated with the same end as the target, but over other sequences the light
was at the opposite end. Hence a good strategy, as expressed by a human, might be:
on the first trial of a sequence, follow the light, and, if the target turned out to be
where the light was, use the target-light correlation to guide the following trials; but
if the target was not associated with the light, turn back to find it, and on the following trials head for the end opposite the light. The problem for the ER method to tackle
is to evolve, from scratch, neural networks that implement some such strategy, based
solely on feedback derived from the time taken to find the target, averaged over many
sequences of trials.

The initial fitness function was indeed based simply on that average; given that the
experimental design implied that half the time the light was correlated with the target
and half the time it was not, there was no easy first-order relationship between the
direction of the light and the direction of what increased the fitness score. As a consequence, evolution fairly soon "discarded" the light sensors as irrelevant (by reducing
their connection weights to zero) and focused on the best simple strategy. This was as
if the robot was following this rule: "first, head to one end at random, and if the target
was not there, then backtrack to the other end." Though this was a reasonable first-
order strategy, the possibility of improving on this, with a second-order strategy
exploiting the light, was no longer available if the light sensors had in effect been
disconnected. The "feathers" were not available to be exapted for "flying."

So the solution found (Tuci, Quinn, and Harvey 2002) was to put in an artificial
bias to make the light sensors salient. The decision was made to weight more heavily
(by a factor of 3) the score for those trials where the light was next to the target, while
retaining the original score when the light and target were anti-correlated. It is the
average of these scores over many sequences of trials that contribute to selective fitness.
In this way the light becomes of first-order significance for fitness, rather than merely
of second-order significance as a learning cue. As a result of this bootstrapping, a
majority of the subsequent evolutionary runs using this modified fitness function
showed the desired results. First, the robot used the light sensors to go toward the light
in search of the target. If the target was not there, the robot "learned from its mistake,"
headed in the other direction, and in subsequent trials headed away from the light.

TT9: When you know that some environmental feature will be salient at a later, more
evolved stage, exploit your foresight by ensuring it is salient early on.

4.6   Neutral Pathways

When one interprets evolution as finding pathways toward the uplands of the fitness landscape, it is tempting to assume that any pathway that fails to rise in fitness should be ignored. This need not be the case. In natural evolution, many mutational changes are neutral, yet still play a role in facilitating subsequent nonneutral changes. The same may well be true in artificial evolution.

One way of visualizing this is to picture a multistory building with numerous interconnecting corridors and staircases. In order to reach the higher stories, it may be necessary to traverse a level corridor to find an available staircase. Even though walking down the corridor does not immediately increase one's height above ground, and even though there is no gradient to the corridor indicating whether you are moving toward or away from the nearest staircase, it is nevertheless a good strategy to search along these neutral pathways.

Such neutrality is associated with redundancy in a mapping from genotype to phenotype. If different genotypes are associated with one and the same phenotype, or indeed with a class of phenotypes of identical fitness, then this is a many to one mapping. It implies that the genotype search space is larger than the phenotype search space, and sometimes this may be good news. This is particularly the case if and when such neutral pathways provide wormholes, or escape routes in hyperspace, allowing a population to escape from what otherwise might have been a local optimum.

One can create artificial abstract fitness landscapes to illustrate this. The NKp landscape proposed by Barnett (1998), an extension of Kauffman's NK landscapes (Kauffman and Weinberger 1989), is one such example. Barnett demonstrated that by varying the parameter p one could increase or decrease the neutrality of the fitness landscape, without needing to alter the ruggedness. A low-neutrality version, with low value of p, might be rugged enough to have lots of local optima, low hills that can trap a population trying to climb to the uplands. A high-neutrality version of the same landscape, with high value of p, can be just as rugged according to standard measures of ruggedness; yet the neutral corridors, or neutral networks, provide escape routes in this case so that the population no longer gets trapped in local optima. The additional neutral pathways will completely transform the evolutionary dynamics.

Barnett (2001) showed that on versions ("epsilon-correlated") of his abstract fitness landscapes a "netcrawler" can be provably optimal. This is a version of a 1 + 1 genetic algorithm; in other words, the population is in effect of size two. He demonstrated that one can optimize the mutation rate for the best balance between exploration and exploitation. Increasing mutation rates increases the rate at which a population "searches blindly along corridors for possible staircases," but also increases the probability of "falling off" such a neutral network. It is desirable to balance these two factors against each other. It is shown (Barnett 2001) that under certain circumstances the provably optimal mutation rate for an "adaptive" netcrawler is such that a fraction $1/e \cong 37$ percent of mutations are neutral. Hence the mutation rate can be adapted on the fly; if currently more than 63 percent of mutated variants prove to be deleterious, the mutation rate should be decreased, and if less, increased.

Such neutral networks are not confined to artificial abstract fitness landscapes. There is good reason to expect that any high-dimensional evolutionary search space, whether biological or artificial, which has sufficient redundancy in a nonarbitrary genotype to phenotype mapping will have such neutral networks. Schuster and colleagues (Schuster et al. 1994) have pioneered the study of these in RNA fitness landscapes. Gavrilets (1997) has applied a (much more abstract and broader) version of these ideas at the species level to explain the dynamics of biological species on a "holey" fitness landscape. Moving closer to ER, the existence and exploitation of neutral networks in an engineering design problem has been demonstrated by Thompson (Thompson 1997; Harvey and Thompson 1997; Thompson and Layzell 2000).

These experiments were based on evolving the connectivity of a field programmable gate array (FPGA): a chip composed of cells with multiplexers ("muxes") in a square array. For these experiments a 10×10 region of cells was used, and binary genotypes (of length 1800 or more) specified the connectivity between cells, the configurations of muxes in each cell, and other synchronizing features. The real, physical FPGA, as thus genetically specified, was tested on its capacity to discriminate between two kinds of input signals: square waveforms of 1 kHz or 10 kHz. The genetic search space is immense; the fitness landscape depends upon the physical properties of the FPGA, which, since it can operate in asynchronous (unclocked) mode, cannot be reliably simulated by a deterministic simulation. In other words, it is a complex fitness landscape whose properties are largely unknown and noisy, and in this respect shares some aspects with biological fitness landscapes (in contrast to synthetic and abstract deterministic fitness landscapes). However it is still possible to keep accurate records of the evolutionary pathway taken.

One example (Thompson and Layzell 2000) used a $1 + 1$ Evolution Strategy: that is, the population was in effect reduced to size two with the current parent and its mutated copy, very similar to Barnett's netcrawler discussed earlier. If the mutant's fitness is greater or equal to its parent, it then becomes the new parent for the next stage. Hence a unique pathway can be recorded, as contrasted with the moving cloud observed when population sizes are bigger. It can be seen (Thompson 2002) that the path did indeed avoid getting trapped in local optima by "escaping" via neutral networks. Subject only to some noise in the (physical) evaluations, the evolutionary pathway was constrained not to go downhill, and there were extended stages without increase in fitness but with genetic modification—exploration of a neutral network. Analysis of the (minimal) population "walking" along neutral pathways shows that—despite fitness remaining unchanged during this level segment—the jump in fitness at the end of the segment was due to a specific mutation that would not have had that beneficial effect at the beginning of the segment. In other words, the neutral drift along that segment was not wasted; it provided a new genetic environment in which that mutation now had a positive effect.

TT10: If there is redundancy in the genotype-phenotype mapping, so as to open up the possibility of neutral networks, then you shouldn't necessarily consider a period of fitness stasis over many generations to be a waste of time—the population may be "looking for a staircase along the corridor."

TT11: Particularly if there is a lot of neutrality in the evolutionary search space, consider a minimal population such as a 1 + 1 GA. . .

TT12: . . . and then consider whether it is appropriate to adapt the mutation rate according to the $1/e \cong 37$ percent rule (Barnett 2001).

4.7   Are We Nearly There Yet?

The experimenter using ER can often sympathize with the young child on an interminable car journey. Supposing one has done one's best to give the initial population the most favorable starting conditions, one has chosen appropriate stepping-stones to make incremental evolution easier, and encouraged the exploration of potentially fruitful pathways, then how long should one wait before becoming impatient for results?

Under some circumstances it is possible to at least put a lower bound on this waiting time. Here we draw on ideas from Worden (1995) and Haldane (1957), but these are revised for the purposes of a GA using a binary genotype. The simplest way to understand this point is via the game of Twenty Questions.

Suppose that this game is played with an agreed search space containing exactly 1,048,576 ($=2^{20}$) objects. Player_1 secretly chooses the target object, and Player_2 can ask twenty questions to elicit Yes/No answers. The optimum strategy for Player_2 is to choose questions that in effect bisect the search space. If exactly one half of the previously agreed-upon set of objects are bigger than a shoe and one half of the objects are smaller than a shoe, then the answer to the initial question "Is it [the target object] bigger than a shoe?" immediately gives 1 bit of information. A less efficient question, such as "Is it bigger than bus?" is less informative; the challenge of the game for Player_2 is to choose the most appropriate question at each stage.

One can view this process as iterative selection in a population of size 2^{20}. At each stage, and each question, the objects in the less fit half are discarded, while we may assume that the objects in the fitter half are duplicated; hence the population size remains unchanged. This is in effect a simple asexual GA, with no mutation, but a vast population size that initially completely spans the search space, then starts to genetically converge, under selection driven by appropriate questions, toward the target. In GA terms, we may treat this as a case of binary genotypes of length 20—and clearly 20 generations (or rounds of selection) is the absolute minimum length of time required, under optimum questioning strategy. This is assuming that the selection process picks the top half—which is in the ball park of a typical selection pressure used in many GAs; for instance, tournament selection with tournaments of size 2, as used in the Microbial GA (Harvey 2001) does just this. Generalizing, for an n-bit genotype, under these circumstances, the minimal expected time to wait is n generations; this is what Worden (1995) calls the "Speed Limit of Evolution." As with the speed of light, this forms an upper bound only achievable under ideal circumstances.

Moving back from the Twenty Questions scenario to evolution, natural or artificial, one can view each round of selection as a very noisy, imperfect version of "the environment posing a binary question"—survive or die. In artificial evolution we may well start with a random initial population that does not span the whole search space, indeed is typically tiny compared to the search space. This, together with the role of mutations and the various inefficiencies combine to make the expected speed of evolution much slower than the theoretical speed limit. In artificial evolution, even under ideal conditions, one rule of thumb is to allow two orders of magnitude slower. Worden (1995) argues that recombination will not provide any added benefit, though his arguments are unclear and not accepted by many. Nevertheless, these back-of-the-envelope estimates are a starting point for estimating how long evolution might take.

So with an artificial evolution problem, with n-bit genotypes, the rule of thumb suggests one should wait for, say, 100n generations before getting impatient. This assumes the selection pressure mentioned earlier, and that one is waiting to find the unique best point in the search space. If redundancy or other reasons imply that any solution within the top-ranked 2^m of the search space would be acceptable, then the rule of thumb dictates $100(n - m)$ as a plausible waiting time.

This is presented here as a rule of thumb, which seems to bear up reasonably well in practice but is not rigorously proven. Extending beyond binary genotypes to those with real values, as might be the case when evolving parameters for neural networks, is controversial. But on the face of it, if a genotype contains n real values, and it is estimated that each such value needs to be defined to 4-bit, or 8-bit accuracy, then a starting place would be to compare this with binary genotypes of length 4n or 8n, for the purposes of relating this to a "speed limit."

TT13: When evolving binary genotypes of length n, under the circumstances previously indicated, consider waiting 100n generations before becoming impatient.
TT14: When evolving genotypes with real values, consider how many bits would give sufficient accuracy and use this to calculate the speed limit.

4.8 Conclusions

Evolutionary robotics can be pursued for scientific or for engineering motives (Harvey et al. 2005). Apart from such motives it also requires experimental and computational skills, and experience builds up a body of insights. Sometimes this results from theoretical analysis of what should work, but often it comes down to intuitions that have yet to be firmly grounded. ER practice, along with most other technical accomplishments, includes hints and practical knowledge that often does not make it into the textbooks.

This chapter has aimed to get some of these "Travel Tips" onto the record. This is not intended as an exhaustive list, and we do not claim to be covering the most important tips—it is a somewhat random set of postcard views from several ER journeys, and other travelers could have emphasized different views. Planning an ER expedition is challenging, but there is a sense of achievement in finding promising pathways toward one's goals.

References

Auerbach, J. E., and J. C. Bongard. 2011. Evolving monolith robot controllers through incremental shaping. In *New Horizons in Evolutionary Robotics, SCI 341*, ed. S. Docieux, N. Bredèche, J-B. Mouret, 55–65. Berlin, Heidelberg: Springer-Verlag.

Barnett, L. 1998. Ruggedness and neutrality—The NKp family of fitness landscapes. In *Alife VI, Proceedings of the 6th International Conference on Artificial Life*, ed. C. Adami, R. K. Belew, H. Kitano, and C. Taylor, 18–27. Cambridge, MA: MIT Press.

Barnett, L. 2001. Netcrawling—Optimal evolutionary search with neutral networks. In *Proceedings of the 2001 Congress on Evolutionary Computation (CEC2001)*, ed. J-H. Kim, B-T. Zhang, G. Fogel, and I. Kuscu, 30–37. Piscataway, NJ: IEEE Press.

Beer, R. D. 1995a. A dynamical systems perspective on agent-environment interaction. *Artificial Intelligence* 72:173–215.

Beer, R. D. 1995b. On the dynamics of small continuous-time recurrent neural networks. *Adaptive Behavior* 3:469–509.

Beer, R. D., and B. Daniels. 2010. Saturation probabilities of continuous-time sigmoidal networks. ArXiv e-prints 1010.1714. http://arxiv.org/abs/1010.1714 (accessed June 21, 2013).

Brooks, R. A. 1991a. Intelligence without representation. *Artificial Intelligence* 47:139–159.

Brooks, R. A. 1991b. Intelligence without reason. In *Proceedings of the 12th International Conference on Artificial Intelligence (IJCAI-91)*, ed. J. Mylopoulos and R. Reiter, 569–595. San Mateo, CA: Morgan Kauffmann.

Dorigo, M. and M. Colombetti. 1997. *Robot Shaping: An Experiment in Behavior Engineering*. Cambridge, MA: MIT Press.

Funahashi, K., and Y. Nakamura. 1993. Approximation of dynamical systems by continuous time recurrent neural networks. *Neural Networks* 6 (6): 801–806.

Gavrilets, S. 1997. Evolution and speciation on holey adaptive landscapes. *Trends in Ecology & Evolution* 12:307–312.

Haldane, J. B. S. 1957. The cost of natural selection. *Journal of Genetics* 55:511–524.

Harvey, I. 1992. Species adaptation genetic algorithms: A basis for a continuing SAGA. In *Towards a Practice of Autonomous Systems: Proceedings of the 1st European Conference on Artificial Life*, ed. F. J. Varela and P. Bourgine, 346–354. Cambridge, MA: MIT Press.

Harvey, I. 2001. Artificial evolution: A continuing SAGA. In *Evolutionary Robotics: From Intelligent Robots to Artificial Life*, LNCS 2217, ed. T. Gomi, 94–109. Heidelberg: Springer-Verlag.

Harvey, I., and A. Thompson. 1997. Through the labyrinth evolution finds a way: A silicon ridge. In *Proceedings of the First International Conference on Evolvable Systems: From Biology to Hardware (ICES96)*, ed. T. Higuchi, M. Iwata, and L. Weixin, 406–422. Heidelberg: Springer-Verlag.

Harvey, I., E. Di Paolo, R. Wood, M. Quinn, and E. A. Tuci. 2005. Evolutionary robotics: A new scientific tool for studying cognition. *Artificial Life* 11 (1–2): 79–98.

Harvey, I., P. Husbands, and D. Cliff. 1993. Issues in evolutionary robotics. In *From Animals to Animats 2: Proceedings of the 2nd International Conference on Simulation of Adaptive Behavior (SAB92)*, ed. J.-A. Meyer, H. Roitblat, and S. Wilson, 364–373. Cambridge MA: MIT Press.

Kauffman, S. A., and E. D. Weinberger. 1989. The NK model of rugged fitness landscapes and its application to maturation of the immune response. *Journal of Theoretical Biology* 141 (2): 211–245.

Mathayomchan, B., and R. D. Beer. 2002. Center-crossing recurrent neural networks for the evolution of rhythmic behavior. *Neural Computation* 14:2043–2051.

Schuster, P., Fontana, W., Stadler, P. F., and I. L. Hofacker. 1994. From sequences to shapes and back—A case study in RNA secondary structures. *Proceedings of the Royal Society B* 255:279–284.

Thompson, A. 1997. An evolved circuit, intrinsic in silicon, entwined with physics. In *Proceedings of the First International Conference on Evolvable Systems: From Biology to Hardware (ICES96)*, ed. T. Higuchi, M. Iwata, and L. Weixin, 390–405. Berlin: Springer-Verlag.

Thompson, A. 2002. Notes on design through artificial evolution: Opportunities and algorithms. In *Adaptive Computing in Design and Manufacture*, ed. I. C. Parmee, 17–26. Berlin: Springer-Verlag.

Thompson, A., and P. Layzell. 2000. Evolution of robustness in an electronics design. In *Proceedings of the Third International Conference on Evolvable Systems: From Biology to Hardware (ICES2000)*, LNCS 1801, ed. J. Miller, A. Thompson, P. Thomson, and T. Fogarty, 218–228. Heidelberg: Springer-Verlag.

Tuci, E., M. Quinn, and I. Harvey. 2002. An evolutionary ecological approach to the study of learning behaviour using a robot-based model. *Adaptive Behavior* 10 (3/4): 201–221.

Vaughan, E. 2007. The evolution of the omni-directional bipedal robot. Doctoral thesis, University of Sussex.

Vickerstaff, R. J., and E. A. Di Paolo. 2005. Evolving neural models of path integration. *Journal of Experimental Biology* 208:3349–3366.

Worden, R. 1995. A speed limit for evolution. *Journal of Theoretical Biology* 176:137–152.

5 Exploring the Roots of Spatial Cognition in Artificial and Natural Organisms: The Evolutionary Robotics Approach

Orazio Miglino and Michela Ponticorvo

5.1 Introduction

If we had a different physiology we would have a different geometry, not necessarily Euclidean.

The behaviors of spatial orientation that an organism displays result from its capacity for adapting, knowing, and modifying its environment; expressed in one word, spatial orientation behaviors result from its *psychology*. These behaviors can be extremely simple—consider, for example, obstacle avoidance, tropisms, taxis, or random walks—but extremely sophisticated as well—for example, intercontinental migrations, orienting in tangled labyrinths, reaching unapproachable areas.

In different species orienting abilities can be innate or the result of a long learning period in which teachers can be involved. This is the case for many vertebrates. Moreover, an organism can exploit external resources that amplify its exploring capacities; it can rely on others' help and in this case what we observe is a sophisticated collective orienting behavior. An organism can use technological devices as well. Human beings have widely developed these two strategies—namely, either exploring its own capacities or learning new orienting skills—and thanks to well-structured work groups (a crew navigating a boat, for instance) and the continuous improving of technological devices (geographical maps, satellites, compasses, etc.), they have expanded their habitat and can easily orient in skies and seas.

It also is possible to observe orienting behaviors in an apparently paradoxical condition: exploring a world without moving one's body. In the present day a lot of interactions between humans and information and communication technologies (mobile phones, PCs, networks) are achieved using orienting behaviors. The best example is the World Wide Web: the explorer in this pure-knowledge universe *navigates* while keeping her body almost completely still.

Spatial orientation behaviors are the final and observable outcome of a long chain made up by very complex psychobiological states and processes. There is no orienting without perception, learning, memory, motivation, planning, decision making, problem solving, and, in some cases, socialization. Explaining how an organism orients in space requires study of all human and animal cognition dimensions and, for this reason, psychology, and in more recent years anthropology, ethology, neuroscience all consider orientation a very interesting field of study. Moreover spatial orientation behaviors can be observed everywhere in the animal kingdom and this allows us to adopt a comparative approach in studying spatial cognition. Orienting in space is in fact one of the few behaviors that is common to most animal species. It is therefore possible to take into account a common behavior, such as homing behavior, detour behavior, or food searching, in analyzing how insects, vertebrates, or mammals realize it. In this way it is possible to identify differences and similarities at various levels of analysis (anatomical, neural, cognitive, etc.). Indeed, moving in an adequate fashion is strictly connected to animal nature: the word "animal" means exactly something that is animated and can therefore move as a whole in its environment. In different kingdoms we also find much simpler organisms whose primary ability is moving properly, such as the well-known paramecium.

In our opinion this indicates that spatial cognition, in whatever form that allows adequate movement, is a very basic requirement in natural organisms. One can find a very long catalog of spatial abilities by studying the ethological, psychological, and veterinarian literature. But, if we abstract from the capabilities of specific species in specific environments, what can we say about the general mechanisms underlying spatial cognition?

For many years of the last century spatial orienting behavior was studied according to a *behaviorist* paradigm. In this perspective attention was focused on defining association laws between environmental stimuli and observable, quantifiable, behavioral responses. Organisms' internal states—today we could say *psychological* states—were left outside of scientific investigation, as they were not directly observable. Organisms were considered a black box that could be described and explained only by observable behavior. This approach led to undoubted successes by American behaviorist psychologists such as the scientific explication of associative learning (see Thorndike 1911; Skinner 1938; Watson 1913, 1914), but they neglected all the cognitive processes that could not be explained by behavioral observation. The only way to unveil the generative mechanisms of behavior was to enter organisms' "heads." The American psychologist C. E. Tolman (1930, 1948) was the one who opened a breach in these heads, opening the way to the study of cognition as we know it. Tolman was a great innovator and anticipated and influenced many issues of successive research in psychology as well as in other disciplines. He believed that organisms, at least the ones he studied (e.g., rats and humans), in order to orient effectively in their environments, built a mental (cognitive) map of the worlds they were immersed in, as we will describe in more detail later. Moreover he was one of the first psychologists who proposed the use of artificial systems, similar to what we call robots, to be used as a means to express psychological theories (Tolman 1939, and for a modern revisit see Miglino et al. 2007).

Since Tolman's work in the 1930s, many psychologists have argued that animals, in particular mammals, use "cognitive maps" of the environment. For Tolman the cognitive map was a necessity: how would he explain otherwise the results he obtained with rats that had learned the labyrinth configuration without being explicitly reinforced? How was it possible for him to hold the stimulus-response association if studies with rats provided suggestive evidence that laboratory animals systematically explored environments they have never experienced before, detect displaced landmarks, and chose new, efficient paths between familiar locations (Tolman 1930; Poucet 1993)?

From that moment on cognitive map theory became very successful, as the metaphor it proposed was attractive: it is easy to imagine animals that use a map in their brain like humans use maps of a city. The theory's success is evident if we consider that cognitive maps have been ascribed to humans (e.g., Tolman 1948; Péruch, Firaudo, and Garling 1989; Herman, Miller, and Shiraki 1987; Coucelis et al. 1987; Baker 1989; Gärling et al. 1990; Gallistel 1990), to dogs, rats, chimpanzees (e.g., Tolman 1948; O'Keefe and Nadel 1978; Thinus-Blanc 1987; Gallistel 1990; Menzel 1973), to birds (e.g., Wallraff 1974; Gould 1982; Baker 1984; Wiltschko and Wiltschko 1987; Gallistel 1990), and to insects (Gould 1986; Gallistel 1989, 1990; Poucet 1993).

According to this theory, different cues, such as data from internal compasses, the position of the sun, and the arrangement of landmarks, are fused to form a map-like representation of the external world. These maps are enduring, geocentric, and comprehensive. In other words, spatial cognition consists of computations in which the input comes from these representations and the output is behavior.

Tolman can therefore be considered a precursor of the computational and representational approach, which has been mainstream in psychology since the 1950s. Summarizing, this point of view tells us that human beings, and many vertebrates, collect information about their environment through their sensory organs and build a mental representation in their brain about it. It is assumed that reality exists externally to us and what we do is to represent it as a kind of picture in our mind and brain. It is worth emphasizing that many mobile robots nowadays are provided with navigation systems starting from this assumption.

In the last decades many empirical clues both on the behavioral and neurocognitive levels have been collected to support this hypothesis. Many lab experiments have showed that various animal species orient in a way that suggests the presence of a Euclidean cognitive map of their environment. In other words our mind and brain may possess a neurocognitive module that organizes spatial information under formal laws of Euclidean geometry. This hypothesis is supported by neural data as well: in the brains of rats, nonhuman primates, and primates there are neurons (place cells and grid cells) that may act as neural building blocks for cognitive representation of space (O'Keefe and Nadel 1978; Hafting et al. 2005; Fyhn et al. 2004) as we will describe later. In parallel, a wide literature on artificial models of the cognitive (see, for example, a recent review by Cheng 2008), connectionist (Rolls and Treves 1998), and robotics (Burgess Donnett, and O'Keefe 1997; Burgess et al. 1998, 2000) kind has tried to formalize and synthesize the experimental results.

However, in the last years, as often happens in scientific practice, a new point of view has arisen, which is different from the mainstream. In fact, the same experimental results can be explained with an alternative frame of reference with respect to computational and representational approaches. An alternative is the action-based, "embodied" perspective that emerged in the late 1980s and early 1990s. This view suggests that many navigational tasks are based on sensorimotor schemes, in which the key input comes not from a complete representation of the geometry of the environment but from local "affordances"(Gibson 1979) that the animal discovers as it moves around.

According to this point of view organisms do not build maps of the external world, but create their own worlds. If we hold this perspective, the question science must answer is not how an organism represents external reality in its mind and brain, but how a continuously evolving living being immersed, situated in a continuously evolving world, creates an embedded cognitive universe in its biological structures (body, brain, genetic code, etc.).

This position has solid roots and has been proposed in different forms by philosophers such as Dewey (1938), Merleau-Ponty (1945), psychologists such as Piaget (1971), and Gibson (1979) and, in a certain sense by Gestalt theory (Köhler 1929). More recently, this point of view has received attention from psychologists O'Regan and Noë (2001), philosopher Clark (1997), biologist and epistemologist Varela (Maturana and Varela 1980, 1992), and the Italian school of Parisi (Parisi, Nolfi, and Cecconi 1992 and Parisi 1994, just to cite two) that extended it to the study of artificial systems cognition (robots and software-simulated agents).

Our studies in evolutionary robotics (ER) belong to this approach and have tried to make concrete the description of cognition in non-representational terms. Our first works (Nolfi et al. 1994; Miglino, Lund, and Nolfi 1995; Miglino, Nafasi, and Taylor 1996) focused on defining techniques and methodologies that allowed us to train, on an evolutionary scale, small mobile robot populations in the same spatial orientation tasks that are used by psychologists with biological organisms (fish, chicks, corvids, rats, humans, etc.).

Using these techniques we have developed a methodology that can help approach the controversy between representational and action-based approaches. The cognitive map metaphor is still too strong. It is often assumed that representations exist, without trying to verify if this is the case or if they are indeed used for spatial cognition.

In this frame of reference sensory input always comes first and action is just the product of input elaboration. The embodiment frame of reference would require a shift in perspective that puts action in its proper place. But this perspective change is almost impossible to run in animal studies. How can we control the animal action?

We may build artificial agents and prevent them from building representation and observe them. The robots generated by evolutionary robotics are the final outcome of an (artificial) evolutionary process of adaptation. Like animals, they are embodied systems that live and act in physical environments. This is crucial. Natural cognition does not emerge in an abstract brain, isolated from the body and the environment, but rather through active interaction with the world and active extraction of meaningful information from this interaction. ER allows us to simulate these interactions comparing the emergent, adaptive behavior of natural and artificial agents (Nolfi and Floreano 2000) and identifying the principles underlying this behavior.

From the point of view of "embodied cognition" (Clark 1997), evolutionary robotics shows how artificial (and natural) organisms can exploit physics and morphology to achieve solutions that would be difficult to achieve using "rational design."

The fact that robots are physical artifacts means that at least in theory they can reproduce the observed behavior of animals in a specific experimental setup. As "embodied" systems, they can exploit the same physics used by animals, displaying biologically plausible behavior. But in robot experiments, unlike animal experiments, researchers have full control over all key variables. For example, they can evolve robots with little or no "memory" or compare systems governed by different control mechanisms.

This direct comparison between behavioral indexes from animal observation in controlled conditions and artificial organisms has been a relevant part of our work. Each experiment was designed with the ambition to put, ideally, the artificial organism we evolve in the same experimental setting used to study spatial behavior in animal labs. This way we apply to robots the understanding process through comparison that is typical of animal psychology. The robotics systems we evolved were treated not as simplified models of much more complex systems (natural organisms), but as new, now much simpler, artificial organisms. In our opinion these studies are the starting point of a new horizon in evolutionary robotics that will lead the artificial evolved agents to share with natural organisms the same environmental niches and the same tasks, and more generally, face the same challenges. In the following pages we will try to support our thesis with four different but related studies concerning spatial cognition.

Which thesis? As introduced before, ER can be used to test the representational and computational point of view in psychology and potentially to support the action-based one.

To describe this approach we would like to use Piaget's (1971) proposal that knowing is an adaptive function, and all knowledge springs from action. Starting from this he proposes a sophisticated model in which (human) cognition develops through complex mechanisms of self-regulation and negative feedback. According to Piaget, the first stage in children's cognitive development is what he calls the "sensory-motor stage." During this phase, which lasts from birth until about the age of two, children's innate reflexes are gradually substituted by mechanisms allowing them to make more efficient contact with their environment. In Piaget's view, the most elementary units of knowledge are "action-based schemes": nonsymbolic structures that originate in children's actions in the external environment and that mediate their future interactions. These motor programs can be generalized to new situations and tend to coordinate to form wider and wider behavioral units. In this perspective, cognition is an active, "embodied" process. Changes of perspective deriving from children's movements, the morphology of their bodies, and the physics of their sensory organs are just as important as the way in which their brains create and process incoming information.

We think that this general principle can be effectively applied to the understanding of spatial cognition: in the following pages we will show in detail how spatial cognition can emerge with no sensory input, how geometry can be represented without representation, how different environmental information can be merged without resorting to language, and how information can be kept in mind without representation, issues that are relevant in the literature about natural organisms and that we will face with the evolutionary robotics approach.

5.2   A Brief Methodological Note

Evolutionary robotics uses a range of different technical devices and computational tools. We will briefly describe here the devices and tools used in our own experiments, including e-puck and Khepera, two miniature mobile robots. Khepera (Mondada, Franzi, and Ienne 1993) and e-puck (Mondada et al. 2009) are designed and built at the Laboratory of Microprocessors and Interfaces of the Swiss Federal Institute of Technology of Lausanne. The robots are round and have an on-board CPU and can be connected to a host computer using a serial port or Bluetooth.

Their motor system consists of two wheels (one on each side), supported by two rigid pivots in the front and back. The wheels can rotate in both directions. The sensory system includes eight infrared sensors, six on the front and two on the back of the robot. This basic apparatus can be easily expanded with additional components. Turrets with their own processors can provide new sensory information. In some of our experiments, we used robot cameras capable of detecting black or white obstacles at a significant distance. Given that the camera is positioned above the robot, and detects distal stimuli, while the infrared sensors are placed along the robot's circumference and detect proximal stimuli, their combined use provides the robot with two quasi-independent sources of information.

Evolving control systems for large populations of physical robots is time consuming and expensive. To resolve this problem, Nolfi (2000) designed and implemented Evo-Robot, a realistic simulator of the Khepera and e-puck robots that makes it possible to run evolutionary experiments in simulation. EvoRobot accurately models the characteristics of Khepera and its interaction with the environment. In our experiments, we often use a modified version of this software. Control systems evolved with EvoRobot can be downloaded to robots and validated in a physical test environment.

In each of the experiments presented below, the robot control system was modeled by an artificial neural network (ANN). From a functional point of view, we use different kinds of input, hidden and output units. The sensor layer consists of two bias units and of units that receive information from the external world. Bias units are units that are always on (that is, their level of activation is always 1) and receive no information from the outside environment. In some cases we have a timekeepers unit whose activation varies as a function of time. The hidden layer consists of units that elaborate the incoming signal. The output layer consists of two motor units, whose activation is fed to the wheels. In some of the experiments, we use an additional decision unit, which represents the robot's decision whether or not to take a specific action (e.g., digging or recognition) in a specific location. Robot movement is determined by the activation of the output units. Robots are bred using a simple form of genetic algorithm (Belew and Mitchell 1996). Each of the experiments we will report here used the same basic procedure. At the beginning of the breeding process, we generated a first generation of 100 robots. We initialized the robots' control systems with random weights. We then tested their ability to perform the experimental task. Performance was measured using a fitness function related to spatial ability. At the end of the testing session, each robot received a fitness score, measuring its ability on the task. After all robots had been tested, the eighty robots with the lowest fitness score were eliminated (truncation selection). We then produced five clones of each of the remaining twenty robots (asexual reproduction). During cloning, a certain percentage of connection weights were incremented by random values uniformly distributed in the interval [−1,+1]. The new neural control systems were implanted in 100 robot bodies thereby creating a second generation of robots. This cycle of testing/selection/reproduction was iterated until no further improvements in fitness were observed.

The environments we use are derived from the animal behavior literature. We mainly use the so-called open field box, a rectangular enclosure commonly used in animal research where the animal can roam freely. In this open field we can modify the shape, for example it can become square, or we can add landmarks, obstacles, or feature information such as color clues.

As always happens in psychology labs, first we train the evolved robots to solve a certain spatial orientation task and then we first test their behavior and then we study their rudimentary artificial neural networks to understand the emerging cognitive organization.

5.3 Spatial Cognition without Environmental Sensory Information

In this section we describe a paradoxical experiment in which the orienting organisms cannot rely on information on the external environment, but only on its internal state. This condition is not found in nature where at least a minimal external sensory source is always present. This is an extreme and simple form of knowledge that is achieved only through action and with no sensory contribution. The ability we investigate is a very basic one: exploring an open enclosure. We propose these experiments to support our thesis that cognition comes from action in an extreme way that is suppressing the sensory side.

5.3.1 How Can Robots with No Sensory Information Exhibit Spatial Cognition?

A new organism is born and it is blind and deaf. It knows nothing about what is outside, but it tries and it can move. It starts receiving some sensation from the inside, maybe hunger. Standing still is not useful: nothing happens. Maybe something edible can be reached by moving. It starts moving and finds unexplored areas where it can get food. Quite soon the organism behaves as if it knows that its world has edges: every time it touches one of the edges it hurts so it comes back to its starting position. Step by step its behavior improves.

What is this newborn organism doing? Which strategy is it using? The newborn organism has found a way to survive in an unknown environment. Next we will describe two extreme examples of how spatial cognition can operate without using sensory information.

How the "Inner World" Can Build Spatial Cognition

At this moment, while you are reading this chapter, you may see your best friend coming toward you, hear the children playing in the courtyard, or smell the inviting sandwich that the person sitting beside you in the train is eating. Maybe in the same moment you feel that you are hungry.

While seeing a friend, hearing the children, or smelling the sandwich each represent a perturbation coming from the outside, perceiving your biological needs represent a perturbation coming from the inside. Outside and inside are therefore two windows our newborn organism can use and whose interaction determines the organism itself.

In this sense, cognition derives from an autopoietic process as defined by Maturana and Varela (1980). According to these authors, a genotype becomes an organism through active interaction with the external environment: it extracts primary resources from the external environment (water, food, etc.) and it transforms them into tissues, bones, organs, systems, and so on. The organism is the factory of itself. This view of the organism can be extended to the genesis of cognitive structures.

As a result of this conception we can consider living systems are self-reproducing systems that compensate for the perturbations arriving from the external world in order to sustain their organization.

However, they also change as a consequence of environmental stimulation. This view strongly resembles what Piaget (1971) said about human cognitive development being an interaction between assimilation and accommodation. Assimilation and accommodation are two complementary processes of adaptation: assimilation changes the external word to adapt it to the internal world while accommodation changes the internal world to adapt it to the external world. Cognition is the result of both processes. This view emphasizes the role of coupled interactions between organisms and the environment.

The internal vs. external issue is relevant also in robotics research. In this field Tom Ziemke (2005, 2007, 2008) has studied intensively this "nonphysical space" that holds past, present, and future together in an "inner world," a notion that has been first introduced by Hesslow (2002) and developed by Grush (2004). The metaphor is powerful and useful because it makes clear the crucial split between the external world the organism is immersed in and its "internal," private world, which is hidden to other organisms but is fundamental in determining the organism's behavior.

If in order to understand the behavior of organisms it is necessary to consider both their inner world and their external environment, constructing artificial organisms make this possible. The simultaneous consideration of both the internal and external aspects of behavior is relevant also if we wish to study spatial cognition. It is usually assumed that this knowledge is founded upon integrating innate schemes with sensory experience, where sensory experience is provided by environmental stimuli and is received by the organism's sensory apparatus. However, in addition to the external environment the organism's body itself is a precious source of stimuli (Parisi 2004). Examples are internal clocks, proprioception, and signals from the gastroenteric apparatus and the hormonal system. Such stimulation is considered relevant to regulate the organism's behavior but not to build up "knowledge" of the external environment. For example, hunger can motivate an organism to choose a certain action to satisfy this need but it is not useful to construct a representation of the environment in which the organism lives.

The newborn organism we introduced at the beginning of this section has to exploit the role of internal stimulation in building knowledge of the environment in which it lives and from which it receives no stimulation at all. This situation is represented in figure 5.1. Figure 5.1a shows the first stage of our artificial organism without sensory

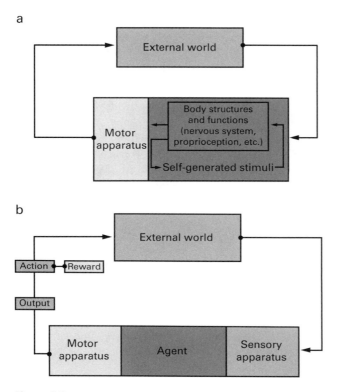

Figure 5.1
Organism/environment dynamic: (a) the organism with internal stimulation only, and (b) the organism with external and internal stimulation.

apparatus but with internal dynamics and action that warrant a link with the world. In a later stage (figure 5.1b) this dynamic is enriched with the introduction of sensory apparatus.

Ponticorvo, Parisi, and Miglino (2009) have considered the issue of introducing sensory apparatus. Let's imagine an organism with a motor apparatus and an internal sensory apparatus, but totally without sensory organs that directly inform the organism concerning the current state of the external environment, just like the newborn organism we described earlier (figure 5.1a). This organism is completely closed inside itself at the sensory level. It can interact with the external environment but it cannot get any direct information from it. The organism is forced to create its own inner world on the basis of purely self-generated stimuli. But the organism cannot be said to be isolated from the external environment in that its actions have effects that modify the physical relation of the organism to the external environment, which the organism can exploit to behave adaptively in the environment.

Can internal stimulation be sufficient to solve a spatial task? Which mechanism do artificial organisms use to adapt? What we have observed is that an adaptive behavior can indeed emerge even in the absence of direct sensory information from the external environment. Even if they are closed in their own self-generated internal world, the simulated robots establish a useful relation with the external environment through their actions. In fact, by realizing and exploiting a precise coordination between produced output and self-generated internal input, in other words, between the external and the internal worlds, the robots are able to successfully adapt to their environment. This is possible because action is accurately selected under evolutionary pressure, and the evolutionary pressure causes the emergence of a kind of resonance between inner world and external world. Through the physical interactions between the organism and the environment, after a demanding search the possibility emerges to utilize action to know the environment, even if there is no sensory input from the environment. In other words, the organism's actions become the vehicle for developing a representation of the environment.

Under evolutionary pressures the agents' neural control architecture extracts and incorporates the statistical regularities and information structure underlying their interactions with the environment, and in this way the external constraints and the internal dynamic resonate. The flow of information between the hidden units and the robot's effectors is actively shaped by the robot's interactions with the environment on an evolutionary scale.

How "Rhythm" Can Build Spatial Cognition

In the previous subsection we have seen that sensor-less artificial organisms could solve effectively some kinds of problems without external stimulation, relying on internal dynamics. These internal dynamics, under evolutionary pressure, were tuned with external dynamic, thus allowing artificial organisms to build spatial cognition. What we observed was that regularities in space were translated into timekeepers in internal dynamics.

In this subsection we would like to add one more piece to the puzzle by discussing the role played by time. Virtually all animals possess "internal clocks" that operate on different timescales, ranging from seconds to years (Farner 1985). These clocks regulate a broad variety of behaviors and body rhythms, from sleep and wakefulness to reproduction and migration. Time and rhythm are, in short, a basic component of animal biology, an issue that was faced in Marocco, Miglino, Walker's (1999) work. They show that adding time sensors to artificial organisms' neural networks allowed these organisms to have a better performance in a wandering behavior, similar to the one already studied in Miglino, Nafasi, and Taylor (1996). The authors compare two kinds of artificial organisms with and without internal clocks. Both were evolved according to evolutionary robotics methods.

Organisms with and without internal clocks were able to adapt to the environment they were evolved in, but organisms with internal clocks proved to be much more efficient in behaving in an environment they had never experienced before. In other words internal clocks help generalization. Using internal timekeepers can significantly simplify a number of cognitive tasks.

In many circumstances is useful to change a behavioral strategy once a certain amount of time has elapsed, regardless of changes in external environment. All animals can rely on internal clocks: it is natural that evolution has used them also for cognitive purposes, to help in building spatial cognition.

Take-home Message

How may the newborn artificial organism explore its world to find food and survive? One possible reply is by exploiting action. Without external sensory information action can become the channel to get in touch with the external world and it can be enough. For this reason these results support our thesis according to which action is cognition's basis.

The behavior of artificial organisms can reflect the particular characteristics of the environment in which the organisms live and are adaptive to, even if they obtain extremely little information from the environment through their sensors, or no information at all. Cognition can emerge from the interaction, made possible by action, between two coupled processes: the agent's internal dynamic and the agent/environment dynamic.

5.4   Spatial Cognition in Natural and Artificial Organisms

In the previous section we have described a paradoxical situation in which an organism is born without external sensory apparatus (figure 5.1a). In this section we take a step further: we consider artificial organisms with sensory apparatus (figure 5.1b) and directly compare them with natural organisms in spatial orientation tasks. This way we show how it is possible to understand the roots of spatial cognition in artificial organisms by focusing on action and not on sensation.

5.4.1   Could Geometry Be Known without an Explicit (Neuro)cognitive Representation of Space?

In this subsection we will describe two experiments in which artificial organisms use geometry without resorting to representation. We will address two specific aspects of geometric knowledge in detail: the use of distance from landmarks and the use of environmental shape.

Landmark Navigation without Representation of Space

Our artificial organism opens its eyes and sees an empty space. "What a bad world I live in!" it may think. Then it moves. Suddenly something interesting appears: a huge black square. Our artificial organism moves toward it. While it is approaching it finds a pink sphere, a yellow cone, and a red cylinder. The artificial organism's world is now populated with relevant cues in proximity of which it can find food, water, or whatever it needs. From now onward it can survive reaching one geometrical figure, or the areas between two of them, or the center of their arrangement.

In the natural world, the role played by these geometric figures is usually assumed by landmarks. Landmarks are distinct features that an animal can recognize in order to orient itself. A landmark can be anything that is easily recognizable such as a monument, a building, or a mountain. Landmarks can be geometric shapes such as rectangles, lines, or circles (as in our artificial imaginary world), and they may include additional information other than the landmark's position. In fact, landmarks have a fixed and known position, relative to which an animal can localize itself. Unsurprisingly, using landmarks to orient is common in animals. Gallistel (1990) has shown that, at least in rats, navigation mechanisms based on landmarks prevails over other navigation mechanisms based on alternative sources of information. Animals can use their distance from a landmark and its bearing to locate their position.

An interesting experiment by Kamil and Jones (1997) shows that some birds can do much more than this. Clark's nutcracker (*Nucifraga columbiana*) is a species of crow with an extraordinary ability to find seeds it has previously hidden. In their experiments, Kamil and Jones demonstrated that in performing this task, the bird could exploit abstract geometric relationships between landmarks. In their work, wild-caught birds were taught to locate a seed, hidden at the midpoint between two landmarks. In the training phase, the seed was partially buried between green and yellow landmarks whose position and distance were changed after each trial. The birds were able to see the seed. In the subsequent, test phase, the seed was completely hidden. In about half the trials, the distance between landmarks was different from the distance in the training phase. This tested the crows' ability to generalize. The results showed that the animals could locate and dig for the seed with a high degree of accuracy, even in conditions they had not experienced during the training phase. Control experiments confirmed that they were exploiting the abstract geometrical relationship between landmarks and did not rely on their sense of smell or on the size of the landmarks. The crows' observed behavior was consistent with Gallistel's hypothesis that orientation uses a Euclidean map of an organism's environment.

However, we would like to show that the action-based approach can also explain these data: landmark navigation does not require explicit representation of space. In the experiment by Miglino and Walker (2004), these findings were replicated, using the ER approach.

The experimental setting in this model reproduced the setting in which Kamil and Jones had conducted their work. Like Kamil and Jones's observation room, robots entered the arena through a "porthole." The arena contained two landmarks (one gray and one black) emulating the green and yellow landmarks in the original experiment. The landmarks were aligned in parallel to the shorter side, with the gray landmark in the northerly position. At the beginning of each trial, the simulated robot was placed at the midpoint of one of the two short sides of the area (always the same side), facing east. During the experiments, the positions of the landmarks and the distances between them were changed after each trial, following the same procedure as in Kamil and Jones's experiments. The robot's sensory input came from a linear camera providing it with a 360-degree field of vision, while the output units consisted of the two motor units plus a decision unit that stopped the robot every time its activation was higher than a defined threshold. This was interpreted as "digging," as explained in the methodological note. Results are shown in figure 5.2.

Results showed that robots succeeded in digging in the correct area, in a way that was comparable with birds. To understand this ability, the authors analyzed the activity of the digging unit, the only unit of the network that was sensitive to the robot's location in the arena in free movement and in imposed position, to understand the mechanism underlying the robot's navigational ability. The data indicated that the robot's ability to exploit geometrical relationships for navigation depended not just on the digging unit but also on the two motor units controlling its movement in the environment.

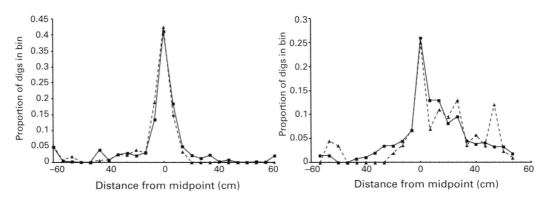

Figure 5.2
Behavioral indexes for robots (on the left) and for birds (on the right). Performance is measured by mean percentage of digs in target area. Squares represent training trials; triangles represent testing results. Robot data from Miglino and Walker 2004; bird data from Kamil and Jones 1997.

In other words, the ability to exploit geometrical relationships between landmarks depended on a process of active perception, with robot action playing a fundamental role. Once again we see how spatial behavior emerges from the embodied interaction between the agent and its environment.

Shape Cognition without Representation of Space

Our artificial organism can move and can calculate distances between landmarks. At this stage, it is ready to learn something more about geometry. Geometry? Does this mean the artificial organism will learn about squares, triangles, rectangles, and other such shapes? Some readers may think that geometry is something we study at school, so how can it be interesting for our young artificial organism? Perhaps some readers who are not geometrically inclined don't want our artificial organism to be bothered with solving complicated geometrical problems.

Indeed, the use of shape for orientation is widespread among animals. Many vertebrates exploit information about the shape of the environment. For example, chimpanzees (Gouteux, Thinus-Blanc, and Vauclair 2001), pigeons (Kelly, Spetch, and Heth 1998), rats (Cheng 1986; Margules and Gallistel 1988), human beings (Hermer and Spelke 1996), newborn chicks (Vallortigara, Zanforlin, and Pasti 1990) can all recognize the geometrical relationships between the walls of a room and use this information as the basis for efficient orientation. It seems clear that many species, and not just *Homo sapiens*, have a detailed geometrical understanding of the portion of the world they inhabit.

Let us consider the experiments run in the experimental setting known as the open field box, described in the methodological note. In open field experiments, rats that have been shown the location of hidden food in a rectangular box are able to navigate toward and dig at that location (or at the rotational equivalent location) in a second, identical box.

During a learning phase, rats learn to localize a visible food patch. Later the rats are tested in a box with the same features as that of the learning box. Here, the rats are to localize a buried food patch put in same position as in the box in the learning phase.

In this experimental condition it was observed that rats systematically produced errors including the "rotational error" (Cheng 1986; Margules and Gallistel 1988). In roughly half of the trials, the rats dug in the correct location. In the other half they dug in the "rotationally equivalent area," that is, the area where they should have dug if the arena had been rotated by 180 degrees. This behavior has been interpreted as evidence that the vertebrate brain contains a "geometry module" that encodes geometric features of the environment, such as distance and direction, and that rats rely exclusively on this information during navigation. In other words, rats understand that the open field box is a rectangle.

These results, about twenty years ago, led Gallistel (1990) to point out that rats have a cerebral module to represent the shape of their environment in terms of Euclidean geometry. In other words, according to Gallistel the vertebrate brain includes a specialized module that provides a Euclidean representation of space. This "geometry module" reconstructs an image of space, which precisely reproduces the quantitative geometrical relationships between objects. In substance, it produces a metric (cognitive) map.

The discovery of "place cells" in the rat hippocampus provided the neural substrate for this hypothesis (O'Keefe and Nadel 1978). Place cells are neurons in the hippocampus whose rate of firing strongly depends on the rat's location in the environment. When the rat is in the area associated with a specific "place field," neurons in the field fire faster than other cells. In O'Keefe's words, a place cell is "a cell which constructs the notion of a place in an environment by connecting together several multi-sensory inputs each of which can be perceived when the animal is in a particular place in the environment" (O'Keefe and Nadel 1978, 425). Complementary information is provided by "head direction cells" (Taube 1998), which fire only when the animal's head is pointing in a specific direction. Recently discovered "grid cells" in dorsocaudal medial entorhinal cortex (dMEC) (Hafting et al. 2005; Fyhn et al. 2004) fire when the animal occupies any one of the vertices of a grid overlaid on the surface of its environment. Together, these findings support the hypothesis that the mammalian brain contains topographic neural maps, representing the spatial environment occupied by the animal.

Place cells and grid cells "fire" when the animal is in a specific location in the environment: a single cell, neurophysiological mechanism supports spatial cognition and geometric information representation.

The geometric module has received notable attention during these last twenty years, but recently this conception has shown some rifts: modularity in geometry is in doubt (Cheng 2008). It seems in fact that geometry could be learned together with other spatial information. This issue, which we will deal with in detail in the next section, is generally approached by assuming that geometry is known with an explicit representation.

But this view is not universally accepted. Other authors argue that animals do *not* use a complete, explicit representation of space. Rather they construct their spatial understanding "on the fly," as they move through space, extracting geometrically meaningful information from the stimuli they receive from the environment. In this view, geometrical knowledge emerges from the interaction between a behaving animal and the environment it inhabits as well as the physical constraints imposed by the environment. In other words, spatial cognition is "situated."

Open field box experiments with rats have provided evidence, which appears to support Gallistel's view, but in 2001, however, Miglino and Lund (2001) used

Table 5.1
Correct identification of target, rotational errors, and misses for rats and for Khepera robots

	Correct	Rotational Errors	Misses
Rats	35	31	33
Khepera	41	41	18

Sources: Data for rats from Margules and Gallistel 1988. Data for Khepera from Miglino and Lund 2001.

techniques from evolutionary robotics to investigate whether robots with no internal representation could produce the same behavior observed in animals. Their aim was to verify whether a construction of a cognitive map of the rectangular box (long vs. short wall) was necessary for obtaining the behaviors described in rats.

They carefully reproduced the experiments using a robot that had no capability of constructing cognitive maps: a Khepera robot governed by an artificial perceptron with direct connection between input and output units. This control system, whose weights were determined via artificial evolution, can make sensorimotor responses, but it cannot build a cognitive map.

The results, reported in table 5.1, showed that with the perceptron control system, the robot was able to navigate to the target in the rectangular box. However, as in the case with rats, the robot would navigate to the rotational equivalent area as many times as to the correct target area. The number of successes was comparable to the ones obtained with rats, and the robots performed fewer misses.

The authors also report a careful analysis of strategy, which underlines that, although the shape of the box (its geometrical characteristics) cannot be seen and represented by the robot, it is assimilated in the robot's behavioral sequences.

This experiment shows that, in the case of open field box settings, similar behavioral indexes can be produced by strategies that do not use a cognitive Euclidean representation (map) of the environmental shape. In fact, the structural characteristics of our robot, which exclude any kind of internal representation, show that the results with rats in open field box experiments cannot be interpreted as evidence of the existence of cognitive maps of the environmental geometry.

A more recent study by Miglino, Ponticorvo, and Bartolomeo (2009) has addressed the question of spatial geometric representation with artificial organisms. In this study, "place units" in robots are produced via artificial evolution. The starting point of this work is a set of studies that show that when animals are restrained, the spatial selectivity of place cells is partially or completely lost. This suggests that the role of place cells in spatial cognition depends not only on the place cells themselves but also on representations of the animal's physical interactions with its environment.

This hypothesis was tested in a population of evolved robots. The results suggest that successful place cognition requires not only the ability to process spatial information but also the ability to select the environmental stimuli to which the agent is exposed. In other words, these results indicate that the action-based perspective can explain data on place cells in spatial cognition.

Take-home Message

If we pronounce the word "geometry" many people will think of a sheet of paper where lines, triangles, cubes, and so forth, are drawn, and of formulas to calculate perimeters, areas, and the like.

In a certain sense this view is not so far from the one derived from the representational hypothesis: if geometry is represented inside the brain in a module, this means that it is represented in an abstract way. On the contrary, the evolutionary robotics experiments we have described in this chapter indicate that geometric information can be used also without resorting to explicit representation. Through action, geometry is exploited by relying on sensorimotor coordination.

5.4.2   How Can Different Types of Spatial Information Be Merged in a Single Neurocognitive System without Language?

The artificial organism, at this stage, can move, distinguish landmarks, and understand the geometrical relationship between them. In the world it lives in and explores, it seems to know that there is a rectangular enclosure with four blue cylinders in the corners and white walls. It often happens, in one of these corners, that something very interesting for our organism appears and disappears fast.

The organism has to localize in which corner the precious reward appears, in order to receive it. Just considering geometry it has a 50 percent chance of success. But one day a new event happens: by a strange coincidence one of the long walls becomes colored. In this case the artificial organism can obtain the desired reward very easily.

The mystery of the colored wall is solved by merging two kinds of information: geometric and nongeometric. Geometric information comes from the difference between a short wall and a long one and sense (left vs. right) whereas the nongeometric information comes from the colored wall.

Not surprisingly, many animals are able to use these forms of information to orient. Evidence suggests that vertebrates orient using geometric and nongeometric feature information. For example, many different animals can locate a region within a larger space by using the distance between their current location and landmarks in the environment. Consider for example rats, birds, fish, primates and human beings. Nongeometric spatial information, such as the color of a landmark, or smells, is also a relevant source.

But, while there is strong evidence that vertebrates know how to exploit geometric and nongeometric information, the exact weight of these sources of information is less clear. In the presence of both cues vertebrates behave in different ways.

Allow us to start from rats, the first species that has undergone the experimental setting of the "Blue Wall" task, an open field box with a colored wall. Rats seem to have some problems in merging geometric and nongeometric information. In the first experiments by Gallistel and colleagues (Gallistel 1989; Margules and Gallistel 1988) rats exploit geometric information in the environment, ignoring contradictory nongeometric information. They therefore often confuse diagonally opposite corners even when featured cues differentiate them. As stated earlier, this was the trigger for the geometric module postulation. Recent results have produced more than one doubt regarding this hypothesis as underlined by the recent review by Cheng (2008). What is observed is that, in some cases, geometric cues were overshadowed by feature cues when spanning entire walls (Pearce et al. 2006).

Other organisms integrate geometric and nongeometric spatial information such as color (Sovrano, Bisazza, and Vallortigara 2002 for fish *Xenotoca eiseni*) or featured cues (Vallortigara, Zanforlin, and Pasti 1990 for young chicks). They use consistently geometric and nongeometric information.

Other puzzling results come from human beings. Systematic rotational errors are made by young children (under five to six years old) in small spaces (Hermer and Spelke 1996), but they use nongeometric information consistently in large spaces or when they grow older (Hermer-Vazquez, Moffet, and Munkholm 2001). Hermer and Spelke (1996), who have studied children's behavior, suggest that merging ability depends on language: mature use of language appears around the age of six, as happens with merging. Moreover the role of language is sustained by results on shadowing findings (Hermer-Vazquez, Spelke, and Katsnelson 1999) where adults engaged in a linguistic task during reorientation ignored a colored wall feature and only used geometric information to reorient. According to these authors, language makes it possible to link geometric and nongeometric information in a single "cognitive representation." If this argument is correct, animals such as rats, which have no superior language functions, are unable to progress beyond "geometric primacy": evolution forces them to rely on the automatic processing provided by the geometric module. Wild-caught mountain chickadees *Poecile gambeli* have shown a nongeometric primacy: "features overshadow geometry" (Gray et al. 2005). It seems therefore that two distinct mechanisms are dedicated to geometric and nongeometric information and these systems do not work together in every condition.

Since Gallistel's (1989) study, according to a certain number of scholars, geometric information (Cheng and Newcombe 2005; Vallortigara, Feruglio, and Sovrano 2005) is coded by a modular system. In fact, Gallistel attributes the ability to orient using geometric cues to a dedicated "brain module" in Fodor's sense of the term (Fodor 1983). In cases of conflict, this geometric module overrides other modules whose output is based on nongeometric information.

Another point of view, supported by behavioral evidence, claims that the modularism is not the only way to approach this controversial issue and language could not play a fundamental role. Newcombe (2002) suggests that geometric and nongeometric information can be integrated. In fact, he claims, "There is no reason to believe that information is encapsulated. It is indeed integrated with other relevant information about the spatial world" (398). This integration "combines input from these various mechanisms in a variable, weighted fashion that reflects characteristics of the input (e.g., size of features or their apparent moveability) and characteristics of the organism's learning history" (Cheng and Newcombe 2005). This hypothesis is supported by studies on monkeys that use nongeometric information if there are large featured cues, but show geometric primacy if these cues are small (Gouteux, Thinus-Blanc, and Vauclair 2001) and by the amazing results involving mountain chickadees *Poecile gambeli* (Gray et al. 2005): when geometric and nongeometric information conflict, there is nongeometric primacy. The authors suggest that this result may be explained by the fact that these wild-caught birds "have little experience with salient right-angle cues."

In a recent paper by Ponticorvo and Miglino (2009), the authors use evolutionary robotics techniques to address the issue of geometric and nongeometric information merging without language. If we describe in operational terms the hypothesis derived from Newcombe 2002, we can say that the variable role of geometric information in different species depends on the frequency with which organisms are exposed to different kinds of spatial information during their adaptive history. In this perspective, language is not a crucial variable to explain geometric primacy, nongeometric primacy, and merging. We assume that any environment contains a certain proportion of geometric information and a certain proportion of nongeometric information. Organisms learn to respond preferentially to the information that is most commonly available to them. Only in a successive phase do they use the less available information. Ponticorvo and Miglino (2009) investigated this hypothesis with ER techniques by manipulating the proportions of geometric and nongeometric information in the environment and observing what patterns of behavior evolve and how information is processed in the artificial brain of evolved robots. Figures 5.3 and 5.4 compare the behavioral indexes for animals and robots.

The results indicate that different orientation abilities can emerge, varying systematically the exposure to different environmental cues. It is possible to evolve agents with different spatial skills by varying the frequency with which they are exposed to different classes of stimuli during their evolution. Agents that evolve in environments providing balanced exposure to geometric and nongeometric cues acquire the ability

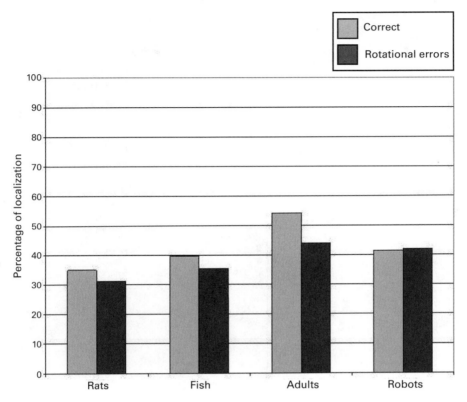

Figure 5.3
Percentage of localization for natural and artificial organisms in the open field box setting. Data for animals are from a review by Cheng and Newcombe (2005), and data on robots are from Miglino and Ponticorvo's unpublished results.

to use both kinds of cue. Agents that are exposed primarily to a single class of cue show primacy.

In more detail, there is a clear interdependency between exposure balance and the ability to integrate geometric and nongeometric information: this ability in fact emerges mainly with balanced exposure to both kinds of information while primacy behaviors emerge in unbalanced exposure conditions.

Moreover it seems that different spatial abilities that emerge in evolved robots are not represented in separate areas in agents' "brains." In fact there is no dissociation between the processing of geometric and nongeometric cues: it is possible to evolve geometric and nongeometric information encoding in the absence of specific modules performing these functions; a modular neural organization is not necessary and

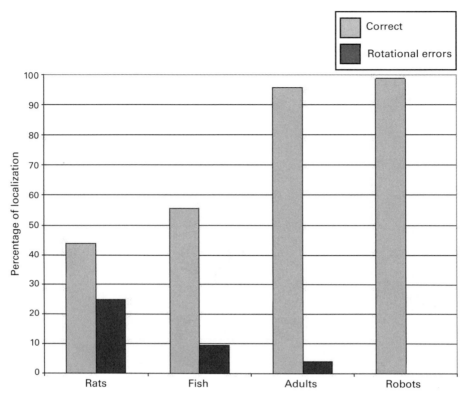

Figure 5.4
Percentage of localization for natural and artificial organisms in the blue-wall task setting. Data for animals are from a review by Cheng and Newcombe (2005), data on robots are from Miglino and Ponticorvo's unpublished results.

therefore language is not necessary to merge different sources of information. Also, on a behavioral level, in spite of various trajectories, adapted to specific environmental conditions, the basic mechanism involved in place cognition appears to be the same for all the agents. The general spatial competence assumes different forms in different environments as different spatial information patterns produce different motor action sequences, but the place recognition mechanism is always the same for all the agents.

Take-home Message
How can the organism find the desired reward? The reply is by merging geometric and nongeometric information, but this merging does not imply the use of language. Different types of spatial information can be merged in single neurocognitive systems

with an appropriate environmental exposition. Shape and colors are put together thanks to the evolutionary pressure that leads to an embedded and situated cognition that allows orientation. Also in this case the key is represented by action: what is needed is not representation but an adequate sensorimotor coordination that links various spatial inputs to the appropriate action.

5.4.3   How Can an External Object Be "Kept in Mind" without an Explicit Representation?

Our artificial organism is now familiar with its environment: it therefore follows quietly the routes to what it is looking for. It now possesses good spatial abilities: it can use landmarks in the world, such as the pink-color sphere; it can use geometric and featured information provided by rectangular squares, circular-arranged walls, or milestones; it can use a time counter to know how much time has elapsed.

At a certain point the organism wants to reach the area where a tall beacon is visible and heads toward it. Suddenly it comes across an enormous obstacle that blocks its path. It is forced to take an indirect route, and it loses sight of the target along the way. In this condition it has to move toward the target but it cannot rely on the beacon mechanism that allows it to approach a target by moving directly toward it.

This kind of behavior is widespread in the animal kingdom. It is not surprising, as many animals have to reach food locations or their nests that are far away from the animal itself, meaning the animal must reach its target even if it is not visible. This behavior, orienting toward a target when it is not visible, is called "detour behavior." Detour behavior is very interesting and it has produced much theoretical debate. In fact, when the target is out of view, there is no stimulus to elicit a response. Consequently many theorists ascribe this behavior to the presence of an internal representation of the target, which allows the animal to maintain its orientation even when the target is out of sight.

In the first years of scientific psychology this ability was considered an example of high-level cognition as it was observed in nonhuman primates such as chimpanzees (Köhler 1925). Moreover, Tolman's (1948) historical study of detour behavior in rats led the famous scholar to postulate the existence of mental spatial maps. In this experiment, Tolman placed rats in a circular labyrinth with a long corridor departing from the center and leading to a reward. The rats easily learned to follow the corridor, but in a later session they found this corridor was blocked, whereas many new alternate corridors now radiated from the circular arena. The rats followed the one that led most directly to the reward, thus demonstrating they remembered where the reward had been. More surprisingly, two-day-old chicks (Regolin, Vallortigara, and Zanforlin 1994) also possess the ability to detour. The chicks were placed in a white cage divided in two by a barrier. The chicks were on one side of the barrier, the target on the other. The chicks' side of the cage contained a corridor. At the end of the corridor, there was a grill that allowed a chick to see the target, which the chick believed to be its mother from the imprinting process. Two openings on the side of the corridor opened onto four compartments, two of which faced away from the target while two faced toward the target. In the experiment, the chick was placed in the corridor, close to the barrier, and allowed to explore its surroundings. Researchers recorded which corners each chick reached and the time they took to reach their target. The results showed that the chicks preferred the compartments facing toward the target. This was taken as evidence that the chicks maintained a mental representation of the target and its position, even when they had lost sight of it along the route.

Walker and Miglino (1999) replicated Regolin and colleagues' experiment in order to understand if an explicit representation was necessary to solve this task. They have replicated this experiment using evolutionary robotics techniques, providing the robots with a control system that precluded the presence of explicit representations of the environment. Each robot was provided with external sensors and with "time sensors" to give the robot a sense of "rhythm."

The authors hypothesized that detour behavior might be derived from other, more primitive forms of exploration and food seeking, namely, the ability to move toward a visible target, to negotiate an obstacle, and to efficiently search an open space. They therefore designed fitness formulae, which rewarded these abilities individually even when they did not lead to successful detours. To elicit robust behavior, the robots were trained in different environments, each consisting of an open field with no external wall. In the first environment, there was no obstacle between the robot and the target. The fitness formula rewarded robots that successfully searched for the target and moved toward it. The second, third, and fourth environments selected for detour behavior. In the training sessions, the robot was always able to perceive the target even when the route to reach it was obstructed. Finally, the authors tested the four best robots from the last generation of each simulation in Regolin and colleagues' apparatus (where the robot could not see the target except at the beginning of the experiment). Behavioral indices, reported in table 5.2, were computed using the same procedures applied by Regolin, Vallortigara, and Zanforlin (1994), that is, by summing up the number of robots choosing specific compartments after a predetermined time.

Table 5.2
Performance of chicks and robots in the experimental apparatus

	Do Not Leave the Corridor	Sec.A	Sec.B	Sec.C	Sec.D	Total
Chicks	5	2	3	9	11	25
Robots	1	0	2	10	11	24

Sources: Regolin, Vallortigara, and Zanforlin 1994; Walker and Miglino 1999.

In this experiment the behavior of the robots successfully replicated that of the chicks. By carefully analyzing the trajectories followed by the robots, Walker and Miglino were able to derive an alternative model of detour behavior, which does not require explicit representation of the target position. In this model detour behavior is built up, step by step, from more primitive behaviors, namely the ability to move toward a target, minimization of the time the robot is out of view of the target, and wall following. These abilities emerge in tight interaction with the environment. Detour behavior is structured by the robot's body and the interaction between the robot body and the environment. It is this interaction that allows the robot to discover new adaptive behaviors: the knowledge resides in the event that links organism and environment.

Take-home Message
In our simulated world the artificial organism can reach the final destination using detour behavior. This ability does not require keeping an explicit representation of the target's position: it is the result of simpler behaviors. Of course, these simpler abilities must emerge in close interaction with the external environment, which is based upon action.

5.5 Conclusions

At the end of this path we have obtained an artificial organism with sophisticated spatial abilities. How did it get these abilities?

The experiments we have described indicate that several forms of spatial behavior can be achieved without internal, symbolic, static representations. The artificial organism's ability to perform specific tasks emerged from the interaction between the robot and its environment: its knowledge springs from the adaptation to environmental constraints. In other words, cognition is not a property of the brain, but a relational property deriving from the organism-environment interaction, mediated by action.

What we have described therefore lends some support to the action-based approach. As we have discussed in the introduction to this chapter, evolutionary robotics is particularly suitable to explore the role of spatial cognition that would otherwise be difficult to address. When we apply ER to a specific problem such as spatial cognition the results provide useful insight into the implicit assumptions of biological and ethological research. The experiments we have described show that, in many cases, there is more than one mechanism capable of generating a particular set of values for this kind of indicator: the representational paradigm is only one possibility for understanding spatial behavior, but the embodied, embedded, situated, implicit one can be powerful as well. In this paradigm action comes first; it is action that determines the input pattern and the resulting next action in an endless chain inscribed in environmental constraints. Therefore knowledge is possible without representation as cognition comes from sensorimotor coordination—in Piaget's words, "a direct action coordination without representation or thinking" (Piaget 1971).

The new horizon in evolutionary robotics that we propose in this chapter consists of using this methodology to discard the representational paradigm in psychology. There is still a lot to be done, however. The challenges we will face in the coming years are various. We would like to introduce them following our imaginary artificial organism as it acquires new spatial abilities step by step. What do we mean by "step by step"? On an evolutionary scale, each step would be a generation, or a hundred generations. On a development scale, each step would a specific period of time or maturation. On a learning scale, each step would be the acquisition of new abilities by trial and error or through supervised teaching. All of these steps should be put together in building artificial organisms that better resemble real-life dynamics.

Until now our artificial organism has only needed to explore its world to get what it needs to survive, but we can imagine that it will want to satisfy new needs. Motivation and emotion are two fundamental dimensions in the animal kingdom and in human life, and introducing them in artificial organisms would probably be difficult but undoubtedly very interesting. At a certain point during its exploration our artificial organism will find another artificial organism, similar or different. It will become a social being that must interact with, understand, anticipate, communicate with, collaborate with, fight against another social being or with numerous social beings. In the animal kingdom sociality is a great trigger for opening up elevated forms of cognition.

Facing these challenges with ER will enable us to develop artificial organisms that behave in complex and unforeseeable ways. They will be a new species, over which we have much more control, to study and analyze in order to solve the puzzling issues of cognition.

References

Baker, R. R. 1984. *Bird Navigation: The Solution of a Mystery*. Sevenoaks: Hodder & Stoughton.

Baker, R. R. 1989. *Human Navigation and Magnetoreception*. Manchester: Manchester University Press.

Belew, R. K., and M. Mitchell, eds. 1996. Adaptive Individuals in Evolving Populations: Models and Algorithms. Redwood City, CA: Addison-Wesley.

Burgess, N., J. G. Donnett, K. J. Jeffery, and J. O'Keefe. 1997. Robotic and neuronal simulation of the hippocampus and rat navigation. *Philosophical Transactions of the Royal Society of London. Series B, Biological Sciences* 352:1535–1543.

Burgess, N., J. G. Donnett, and J. O'Keefe. 1998. Using a mobile robot to test a model of the rat hippocampus. *Connection Science* 10:291–300.

Burgess, N., A. Jackson, T. Hartley, and J. O'Keefe. 2000. Predictions derived from modelling the hippocampal role in navigation. *Biological Cybernetics* 83:301–312.

Cheng, K. 1986. A purely geometric module in the rat's spatial representation. *Cognition* 23:149–178.

Cheng, K. 2008. Whither geometry? Troubles of the geometric module. *Trends in Cognitive Sciences* 12:355–361.

Cheng, K., and N. S. Newcombe. 2005. Is there a geometric module for spatial orientation? Squaring theory and evidence. *Psychonomic Bulletin & Review* 12:1–23.

Clark, A. 1997. Being There: Putting Brain, Body, and World Together Again. Cambridge, MA: MIT Press.

Coucelis, H., R. G. Golledge, N. Gale, and W. Tobler. 1987. Exploring the anchor-point hypothesis of spatial cognition. *Journal of Environmental Psychology* 7:99–122.

Dewey, J. 1938. *Experience and Education*. New York: Macmillan.

Farner, D. S. 1985. Annual rhythms. *Annual Review of Physiology* 47:65–82.

Fodor, J. A. 1983. *The Modularity of Mind. An Essay on Faculty Psychology*. Cambridge, MA: MIT Press.

Fyhn, M., S. Molden, M. P. Witter, E. I. Moser, and M.-B. Moser. 2004. Spatial representation in the entorhinal cortex. *Science* 305:1258–1264.

Gallistel, C. R. 1989. Animal cognition, the representation of space, time and number. *Review of Psychology* 40:155–189.

Gallistel, C. R. 1990. *The Organization of Learning*. Cambridge, MA: MIT Press.

Gärling, T., A. Böök, E. Lindberg, and C. Arce. 1990. Is elevation encoded in cognitive maps? *Journal of Environmental Psychology* 10: 341–351.

Gibson, J. J. 1979. *The Ecological Approach to Visual Perception*. Boston: Houghton Mifflin.

Gould, J. L. 1982. The map sense of pigeons. *Nature* 296:205–211.

Gould, J. L. 1986. The local map of honey-bee: Do insects have cognitive maps? *Science* 232:861–863.

Gouteux, S., C. Thinus-Blanc, and J. Vauclair. 2001. Rhesus monkeys use geometric and nongeometric information during a reorientation task. *Journal of Experimental Psychology* 130:505–509.

Gray, E. R., L. L. Bloomfield, A. Ferrey, M. L. Spetch, and C. B. Sturdy. 2005. Spatial encoding in mountain chickadees: Features overshadow geometry. *Biology Letters* 1:314–317.

Grush, R. 2004. The emulation theory of representation: Motor control, imagery, and perception. *Behavioral and Brain Sciences* 27:377–442.

Hafting, T., M. Fyhn, S. Molden, M.-B. Moser, and E. I. Moser. 2005. Microstructure of a spatial map in the entorhinal cortex. *Nature* 436:801–806.

Herman, J. F., B. S. Miller, and J. H. Shiraki. 1987. The influence of affective associations on the development of cognitive maps of large environments. *Journal of Environmental Psychology* 7:89–98.

Hermer, L., and E. S. Spelke. 1996. Modularity and development: The case of spatial reorientation. *Cognition* 61:195–232.

Hermer-Vazquez, L., A. Moffet, and P. Munkholm. 2001. Language, space, and the development of cognitive flexibility in humans: the case of two spatial memory tasks. *Cognition* 79:263–299.

Hermer-Vazquez, L., E. Spelke, and A. Katsnelson. 1999. Sources of flexibility in human cognition: Dual task studies of space and language. *Cognitive Psychology* 39:3–36.

Hesslow, G. 2002. Conscious thought as simulation of behaviour and perception. *Trends in Cognitive Sciences* 6 (6): 242–247.

Kamil, A. C., and J. E. Jones. 1997. Clark's nutcrackers learn geometric relationships among landmarks. *Nature* 390:276–279.

Kelly, D. M., M. L. Spetch, and C. D. Heth. 1998. Pigeons' (*Columbia livia*) encoding of geometric and featural properties of a spatial environment. *Journal of Comparative Psychology* 112:259–269.

Köhler, W. 1925. *The Mentality of Apes*. New York: Harcourt Brace.

Köhler, W. 1929. *Gestalt Psychology*. New York: Liveright.

Margules, J., and C. R. Gallistel. 1988. Heading in the rat: Determination by environmental shape. *Animal Learning & Behavior* 16:404–410.

Marocco, D., O. Miglino, and R. Walker. 2000. Do it with rhythm—How internal clocks can simplify life in artificial and biological organisms. In *Proceedings of the International Interdisciplinary Workshop on Studies on the Structure of Time: From Physics to Psycho(patho)logy*, ed. R. Buccheri, V. Di Gesù, and M. Saniga, 107–121. New York: Kluwer Academic/Plenum Publishers.

Maturana, H. R., and F. J. Varela. 1980. *Autopoiesis and Cognition*. Boston: Reidel.

Maturana, H. R., and F. J. Varela. 1992. The tree of knowledge. In *The Biological Roots of Human Understanding*, 242. Boston: Shambhala.

Menzel, E. W. 1973. Chimpanzee spatial memory organization. *Science* 182:943–945.

Merleau-Ponty, M. 1945. *Phenomenologie de la perception*. Paris: Gallimard.

Miglino, O., O. Gigliotta, M. Cardaci, and M. Ponticorvo. 2007. Artificial organisms as tools for the development of psychological theory: Tolman's lesson. *Cognitive Processing* 8 (1): 261–277.

Miglino, O., and H. H. Lund. 2001. Do rats need Euclidean cognitive maps of the environmental shape? *Cognitive Processing* 4:1–9.

Miglino, O., H. H. Lund, and S. Nolfi. 1995. Evolving mobile robots in simulated and real environments. *Artificial Life* 2 (4): 417–434.

Miglino, O., K. Nafasi, and C. E. Taylor. 1996. Selection for wandering behaviour in a small robot. *Artificial Life* 2 (1): 101–116.

Miglino, O., M. Ponticorvo, and P. Bartolomeo. 2009. Place cognition and active perception: A study with evolved robots. *Connection Science*, 21 (1): 3–14.

Miglino, O., and R. Walker. 2004. An action-based mechanism for the interpretation of biometrical clues during navigation. *Connection Science* 16 (4): 267–281.

Mondada, F., M. Bonani, X. Raemy, J. Pugh, C. Cianci, A. Klaptocz, S. Magnenat, J.-C. Zufferey, D. Floreano, and A. Martinoli. 2009. The e-puck, a robot designed for education in engineering. *Proceedings of the 9th Conference on Autonomous Robot Systems and Competitions* 1 (1): 59–65.

Mondada, F., E. Franzi, and P. Ienne. 1993. Mobile robot miniaturization: A tool for investigation in control algorithms. In *Proceedings of the Third International Symposium on Experimental Robotics*, ed. T. Yoshikawa and F. Miyazaki, 501–513. Tokyo: Springer-Verlag.

Newcombe, N. 2002. The nativist-empiricist controversy in the context of recent research on spatial and quantitative development. *Psychological Science* 13:395–401.

Nolfi, S. 2000. *EvoRobot 1.1 User Manual*. Rome: Institute of Psychology, CNR.

Nolfi, S., and D. Floreano. 2000. *Evolutionary Robotics: The Biology, Intelligence, and Technology of Self-Organizing Machines (Intelligent Robotics and Autonomous Agents)*. Cambridge, MA: MIT Press.

Nolfi, S., D. Floreano, O. Miglino, and F. Mondada. 1994. How to evolve autonomous robots: Different approaches in evolutionary robotics. In *Proceedings of the International Conference Artificial Life IV*, ed. R. Brooks and P. Maes, 190–197. Cambridge, MA: MIT Press.

O'Regan, J. K., and A. Noë. 2001. A sensorimotor account of vision and visual consciousness. *Behavioral and Brain Sciences* 24 (5): 939–1031.

O'Keefe, J., and L. Nadel. 1978. *The Hippocampus as a Cognitive Map*. Oxford: Clarendon.

Parisi, D. 1994. Are neural networks necessarily passive receivers of input? In *Neural Networks in Biomedicine*, ed. F. Masulli, P. G. Morasso, and A. Schenone, 113–124. Singapore: World Scientific.

Parisi, D. 2004. Internal Robotics. *Connection Science* 16 (4): 325–338.

Parisi, D., S. Nolfi, and F. Cecconi. 1992. Learning, behaviour, and evolution. In *Toward a Practice of Autonomous Systems*, ed. F. Varela and P. Bourgine, 207–216. Cambridge, MA: MIT Press.

Pearce, J. M., M. Graham, M. A. Good, P. M. Jones, and A. McGregor. 2006. Potentiation, over-shadowing, and blocking of spatial learning based on the shape of the environment. *Journal of Experimental Psychology. Animal Behavior Processes* 32:201–214.

Péruch, P., K. D. Firaudo, and T. Garling. 1989. Distance cognition by taxi drivers and the general public. *Journal of Environmental Psychology* 9: 233–239.

Piaget, J. 1971. *Biology and Knowledge: An Essay on the Relations between Organic Regulations and Cognitive Processes.* Chicago: University of Chicago Press.

Ponticorvo, M., and O. Miglino. 2009. Encoding geometric and nongeometric information: a study with evolved agents. *Animal Cognition* 13 (1): 157–174. doi:10.1007/s10071-009-0255-7.

Ponticorvo, M., D. Parisi, and O. Miglino. 2009. The autopoietic nature of the "inner world": A study with evolved "blind" robots. In *ABiALS 2008*, LNAI 5499, ed. G. Pezzulo, M. V. Butz, O. Sigaud, and G. Baldassarre, 115–131. Berlin: Springer..

Poucet, B. 1993. Spatial cognitive maps in animals: New hypotheses on their structure and neural mechanisms. *Psychological Review* 100:163–182.

Regolin, L., G. Vallortigara, and M. Zanforlin. 1994. Object and spatial representations in detour problems by chicks. *Animal Behaviour* 48:1–5.

Rolls, E. T., and A. Treves. 1998. *Neural Networks and Brain Function.* Oxford: Oxford University Press.

Skinner, B. F. 1938. *The Behaviour of Organisms: An Experimental Analysis.* New York: Appleton-Century-Crofts.

Sovrano, V. A., A. Bisazza, and G. Vallortigara. 2002. Modularity and spatial reorientation in a simple mind: Encoding of geometric and nongeometric properties of spatial environment by fish. *Cognition* 85:51–59.

Taube, J. S. 1998. Head direction cells and the neurophysiological basis for a sense of direction. *Progress in Neurobiology* 55:225–256.

Thinus-Blanc, C. 1987. The cognitive map concept and its consequences. In *Cognitive Processes in Animal and Man*, ed. P. Ellen and C. Thinus-Blanc, 1–19. The Hague: Martinus Nijhoff.

Thorndike, E. L. 1911. *Animal Intelligence.* New York: Macmillan.

Tolman, E. C. 1930. Insight in rats. *Psychology* 4:215–232, 277–278.

Tolman, E. C. 1939. Prediction of vicarious trial and error by means of the schematic sowbug. *Psychological Review* 46:318–336.

Tolman, E. C. 1948. Cognitive maps in rats and men. *Psychological Review* 36:13–24.

Vallortigara, G., M. Feruglio, and V. A. Sovrano. 2005. Reorientation by geometric and land-mark information in environments of different spatial scale. *Developmental Science* 5 (8): 393–401.

Vallortigara, G., M. Zanforlin, and G. Pasti. 1990. Geometric modules in animal spatial representations: A test with chicks (*Gallus gallus*). *Journal of Comparative Psychology* 104:248–254.

Walker, R., and O. Miglino. 1999. Replicating experiments in "detour behaviour" with artificially evolved robots: An A-Life approach to comparative psychology. In *Proceedings of the 5th European Conference on Artificial Life (ECAL99)*, 205–214. London: Springer-Verlag.

Wallraff, H. G. 1974. *Das Navigationssystem der Vögel. Ein theoretischer Beitrag zur Analyse ungeklärter Orientierungsleistungen*. München: Schriftenreihe Kybernetik, Oldenbourg.

Watson, J. B. 1913. Psychology as the behaviorist views it. *Psychological Review* 20:158–177.

Watson, J. B. 1914. *Behavior: An Introduction to Comparative Psychology*. New York: Holt.

Wiltschko, W., and R. Wiltschko. 1987. Cognitive maps and navigation in homing pigeons. In *Cognitive Processes and Spatial Orientation in Animal and Man I*, ed. P. Ellen and C. Thinus-Blanc, 201–216. The Hague: Martinus Nijhoff.

Ziemke, T. 2005. Cybernetics and embodied cognition: On the construction of realities in organisms and robots. *Kybernetes* 34 (1/2): 118–128.

Ziemke, T. 2007. The embodied self—Theories, hunches and robot models. *Journal of Consciousness Studies* 14 (7):167–179.

Ziemke, T. 2008. On the role of emotion in biological and robotic autonomy. *Bio Systems* 91:401–408.

6 Why Morphology Matters

Josh Bongard

One can distinguish between traditional and evolutionary robotics (ER) by the way in which each community generates controllers: traditional roboticists hand-design or use learning methods to create control policies, while evolutionary roboticists employ evolutionary algorithms. What further distinguishes these two approaches is that evolutionary algorithms may also be used to optimize robot morphology as well as the control policy. This chapter traces the history of this practice and outlines how we as a community are transitioning from questions regarding *how* to evolve morphology to *why* one should do so. Here I outline seven such reasons: selecting or evolving an appropriate morphology can (1) simplify control, (2) make seemingly difficult tasks easier, (3) increase evolvability, (4) provide new behaviors, (5) facilitate the extraction of information from the environment, (6) generate new research questions, and (7) improve scalability.

6.1 Introduction

Embodied cognition was an intellectual rebellion that entered the field of artificial intelligence (AI) in the early 1990s (Brooks 1991a, 1991b), and challenged the prevailing (and still majority) view in AI that intelligence can be replicated in a computer without requiring interaction with the external world (Minsky 1974; Bechtel 1990). Brooks outlined a strong view in which internal modeling was not required for realizing relatively sophisticated behavior such as locomotion over uneven terrain (Brooks 1986) and, later, social interaction (Brooks et al. 1999). Indeed internal processing was minimized as much as possible; instead, emphasis was placed on exploiting the situated and embodied nature of the robot.

This movement has steadily been gaining ground since that time (Clark 1996; Pfeifer 1999; Pfeifer and Bongard 2006), but a large majority of the AI and even robotics community continue to focus extensively on the control side of behavior realization, with minimum attention paid to the robot's morphology. One reason for this emphasis may be that AI and robotics grew out of cybernetics, which in turn was founded on control theory. Furthermore, the formal, mathematical traditions of the field promote approaches that provide guarantees of convergence or the discovery of optimal solutions, or both. Such guarantees are very difficult or impossible in stochastic optimization processes such as evolutionary algorithms, much less when the optimization process is extended to the morphology of the robot. It may be that this contributes to why evolutionary robotics is not a popular approach among many mainstream robotics and AI practitioners.

Nature however provides abundant examples of organisms with diverse body plans, and an ever-greater diversity of adaptive mechanisms by which those body plans support behavior. Biorobotics (Webb and Consi 2001), a sister field of evolutionary robotics, has demonstrated several successes of replicating both the control and morphological adaptations of individual animals in machine form such that the machines also demonstrate one or more of the animal's behaviors. However, biorobboticists tend to copy the actual body plan of animals, rather than copy the evolutionary mechanisms that produced that body plan in the first place, as is done in evolutionary robotics.

If the right morphology can indeed facilitate behavior, the question then arises as to how to select an appropriate machine body plan for the task environment and desired behavior. A prevailing view in robotics is that choosing such a body plan is much more intuitive that designing controllers: "Humans are much better at designing physical systems than they are at designing intelligent control systems: complex powered machines have been in existence for over 150 years, whereas it is safe to say that no truly intelligent autonomous machine has ever been built by a human" (Nelson, Barlow, and Doitsidis 2009, 22). However, a growing number of examples surveyed in this chapter illustrate that selecting an appropriate body plan is rarely an intuitive process, and that therefore automating the selection process using evolutionary computation may indeed allow for the realization of increasingly intelligent and autonomous machines.

6.1.1   The Role of Morphology in Animal Behavior

As already mentioned, many of the geometric layouts, material properties, and mechanical mechanisms of animals' body plans are shaped by evolution to support particular behaviors. A well-studied example in robotics is human bipedal locomotion. Human legs are structured to support extremely energy-efficient travel by exploiting the passive swing of the leg and loading of the ankle, and therefore support long-distance travel. Indeed several researchers have been able to reproduce this energy-efficient gait in robots through careful hand-tuning of the robot's morphology (McGeer 1990; Collins et al. 2005). However, for many morphological adaptations it is difficult to determine which behaviors they evolved to support. Whether the quadrupedality seen in higher animals or the hexapody seen in insects is a result of historical accident or serves adaptive behaviors is unknown, and there exist a panoply of hypotheses for why bipedality evolved in early humans (Lovejoy 1980; Morgan 1982; Jablonski and Chaplin 1993; Hunt 1996).

For this reason it is difficult to determine which aspects of an organism's morphology to replicate in a robot, as some aspects may be dictated by physiological constraint or are the result of historical accident. An example of this former constraint is that it seems likely that bipedalism was a simpler path for evolution to take to free up the upper extremities rather than evolving a new pair of limbs, as the quadrupedal body plan is an extremely conserved trait across the animal kingdom. Therefore, although instantiating bipedalism in robots may be desirable so that they can operate in a world built for human body plans, if not done correctly, a legged robot may not be capable of the behavior that legs originally evolved for: energy-efficient transport. Similarly, whether to implement four or five fingers on a robot hand is probably a less important design decision than whether to include an opposable digit, which supports a wider range of grasping strategies than a hand without such a digit (Wilson 1998).

As these two examples illustrate, choosing an appropriate body plan for a robot is not so easy a task as it seems. An alternative strategy therefore is to replicate the evolutionary mechanism that produced the body plan originally: biological evolution. Artificial evolution may then discover a body plan that, while possibly not similar to any found in nature, is well suited to the task environment, desired behavior, and control policy evolved along with it.

6.1.2 "How" versus "Why"

Karl Sims was the first to demonstrate that evolutionary computation could be used to evolve both the morphology and control policies for autonomous agents in a physics-based simulator (Sims 1994). He obtained many agents with morphologies that were and were not biologically familiar. He employed an evolutionary algorithm with a genotype-to-phenotype mapping that included recursion and therefore tended to produce body plans with repeated segments.

This work stimulated a subsequent wave of research that explored different ways of evolving morphology. One research line employed direct encoding schemes (Ventrella 1994; Lipson and Pollack 2000), while another line expanded the recursive Sims-type mapping by incorporating Lindenmayer systems (Prusinkiewicz and Lindenmayer 1990) into the encoding scheme (Hornby and Pollack 2001). Others still pursued more biologically realistic encodings that simulated genetic regulatory networks to grow the body plan (Eggenberger 1997), and later both the body plan and control policy (Bongard 2002) of simulated agents. Examples of robots evolved with this latter approach are shown in figure 6.1q–u.

Yet most of this work focused on *how* to evolve morphology, rather than *why* one should do so: it is imperative for the field of evolutionary robotics to accumulate such

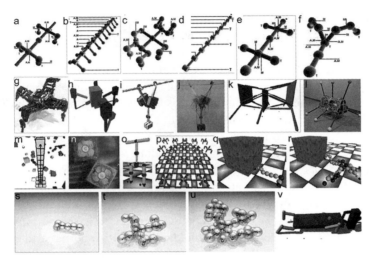

Figure 6.1
A selection of the author's past projects involving embodied cognition. (a–f) Simulated machines
with differing morphologies, yet identical sensor, motor, and control systems were evolved to
isolate the effect of morphology on evolvability (Bongard and Pfeifer 2002). A physical autonomous machine (g) was reported in Bongard, Zykov, and Lipson 2006 that is capable of creating
a simulation of its own morphology (h). The simulation of a brachiating machine (i) was used
to prototype passive dynamic behaviors before instantiating them in a physical machine
(j) (Frutiger, Bongard, and Iida et al. 2002). The simulation of a machine with both serial and
parallel actuated linkages (k) was used to prototype nonintuitive control strategies that were later
used on the physical version of the machine (l) (Zykov, Bongard, and Lipson 2004). A simulation
framework for prototyping stochastic self-assembling machine modules was developed, and used
to discover assembly plans for producing three-dimensional structures such as the tower shown
in (m). Results from the simulation were used to build a macroscale physical prototype of this
technology (n) (White et al. 2005). (o) Both mass distribution and control parameters of a simulated bipedal machine were evolved to produce locomotion in Bongard and Paul 2001. (p) Strategies by which multiple machines may share self-models and controllers were investigated in
Bongard 2009. (q–u) Evolving the body plans and control circuits for simulated machines. (q,r):
Evolution of block pushing behavior (Bongard and Pfeifer 2001). (s–u) Illustrations indicating the
combined growth of the body plan during ontogenesis as well as the internal sensor, motor, and
control circuits (internal networks). (v) A virtual robot capable of both legged locomotion and
object manipulation (Auerbach and Bongard 2009b).

reasons in order to justify the added complexity of these methods. This chapter will explore seven reasons why a machine's body plan should be carefully hand- or automatically designed, beyond the one already mentioned:[1]

1. Choosing or evolving an appropriate body plan can simplify control compared to when an inappropriate body plan is employed—*morphology simplifies control*;
2. Seemingly difficult tasks, approached with purely computational methods, become easier if the methods incorporate an appropriate morphology—*morphology eases tasks*;
3. Although incorporating morphological parameters into the evolutionary process increases the dimensionality of the search space, doing so can often improve the probability of finding useful behavior—*morphology increases evolvability*;
4. By exploring the space of robots in which the topology of the robot's body plan may change during behavior, new behaviors not available with a fixed morphology can be realized—*morphology affords new behaviors*;
5. With the right body, a robot can use it to systematically extract useful information from the environment—*morphology supports self-exploration*;
6. With the adoption of certain body plans, research questions not yet explored in artificial intelligence arise—*morphology creates new research questions*; and
7. What constitutes a "good" morphology becomes less intuitive as the complexity of the task increases—*morphology supports scalability*.

6.2 Morphology Simplifies Control

The task environment in which a robot must act dictates much of the control policy that the robot should adopt in order to perform its task successfully. However, the task environment also dictates what kinds of morphology are appropriate: a poorly chosen morphology will require a more complex control policy than a well-chosen morphology. As a simple example, consider a robot that should move forward over flat terrain. Clearly a wheeled robot will require simpler control than a legged robot: a wheeled robot may simply supply constant torque to all of its wheels while a legged robot must orchestrate the motion of its legs. However, for more complex task environments, it is not always so clear how to devise a morphology that will simplify control. Therefore, a growing number of examples have been put forward in the robotics and evolutionary robotics community. One such example, passive dynamic walking machines, has already been mentioned, in which little or no control is required if the body is designed properly. However, it is often the case that human intuition or examples from nature fail to suggest what an appropriate robot body plan should be for a given task. In such circumstances evolutionary algorithms can be used to discover an appropriate body plan for the task at hand.

Lichtensteiger and Eggenberger (1999) described a mobile robot that simulates the facets of an insect's eye, and demonstrated that evolving the distribution of these facets on the robot (which resulted in nonuniform distributions) simplified the visual processing required to calculate the time at which the robot would contact an external object.

The author developed a method that combines evolution and ontogeny: virtual organisms are grown in the environment in which they must behave, and evolution in turn shapes the form of these growth programs (Bongard and Pfeifer 2001). This was accomplished by simulating genetic regulatory networks (Eggenberger 1997): genomes that are evolutionarily modified may contain noncoding and coding regions. These coding regions are treated as genes, and the parameters within these regions dictate the gene's behavior. Genes produce simulated substances that diffuse through the body of the robot as it grows in its environment. Some of these substances may affect the expression of other genes near the point of the substance's production or, through passive diffusion, the substance may affect gene expression in distal parts of the robot.

Other substances may cause phenotypic change. The robots are composed of spherical modules (figure 6.1q–u), and modules may grow and split in response to substance concentration. The modules also contain neurons and connecting synapses (visualized by the internal networks in figure 6.1s–u); substances may create, destroy, or move neurons throughout the robot's body, and may cause synapses to grow from one neuron, through the body, and attach to other neurons. Additional synapses may connect sensors to neurons, neurons to motors, or sensors directly to motors. Using this approach, few assumptions are made about the form of the robots' body plans and neural controllers, and evolution is free to discover an appropriate body plan and neural controller for the task at hand.

In one experiment in which robots were selected for approaching and pushing large objects in their environment (figure 6.1q,r), two robots were observed to have long front appendages and exhibited a wave-like form of locomotion reminiscent of that seen in inchworms. Both robots were evolutionarily related, and were found to possess very similar neural patterning as illustrated in figure 6.2.

A human engineer asked to design a controller for such a body plan would most likely favor a centralized architecture in which a central timer orchestrated a sequence of motions along the appendage's length. However, artificial evolution here discovered a simpler solution that does not require timing or orchestration, but rather exploits the interactions between the robot's body plan (a series of spherical modules) and the environment (gravity pulls lifted modules back to the ground) and controller (distributed direct sensor-motor reflexes).

This example points to one of the fundamental differences between more formal learning methods and evolutionary algorithms: the former are suited for parametric

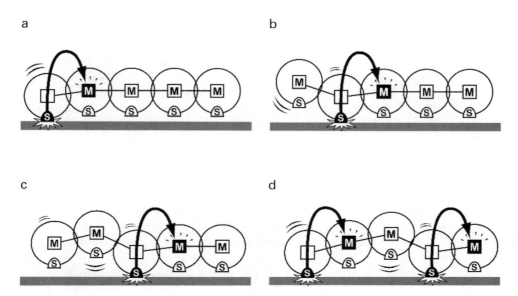

Figure 6.2

The evolved morphology, control policy, and resultant action of the front appendages of the robots shown in figure 6.1q,r. The front appendage is comprised of a series of morphological modules (large circles) attached to each other by one degree-of-freedom rotational joints. Each joint is motorized (M). The evolved genetic regulatory network places a touch sensor (S) in each module, and connects it by a synapse (arrows) to the motorized joint in the neighboring module. Gravity and the mass distribution of the appendage causes the distal tip to contact the ground, which causes the touch sensor to fire and activate the motor it connects to. This causes the joint to rotate the tip off the ground, thereby causing the second module to contact the ground. This in turn activates the next touch sensor and its associated motor, and so on, producing a wave of motion along the appendage's length without requiring centralized control (Courtesy MIT Press).

optimization in which guarantees of convergence are required, while the latter allow both structural (i.e., topological) and parametric improvement in which such guarantees are more difficult to provide. Currently, popular learning methods support that there are a fixed number of parameters that must be optimized: these parameters may specify the parameters of a controller with a fixed architecture, or aspects of the robot's morphology. However, many evolutionary algorithms do not presuppose the dimensionality of candidate solutions. This is very much the case in indirect genotype-to-phenotype mappings, in which there is dissociation between the complexities of the genotype and phenotype. Indeed the more complex robot shown in figure 6.1r is a descendent of the simpler robot in figure 6.1q, although both were grown from a genotype with the same number of genes (Bongard and Pfeifer 2001).

6.3   Morphology Eases Tasks

One of the surprising results accumulating in the embodied cognition literature is that seemingly challenging tasks when tackled with disembodied algorithms are rendered much simpler when an intelligent agent is allowed to interact with the environment using its body. A particularly striking example comes from the work of Metta and Fitzpatrick (2003), in which a humanoid robot is tasked with object segmentation. This task has primarily been cast as a purely perceptual problem in the computer vision literature in which objects in a cluttered scene should be visually separated ("segmented") from the background and other objects.

Metta and Fitzpatrick describe a series of experiments in which a humanoid robot moves its arm, and observes the resulting motion in its visual field. When the arm comes in contact with various objects, the robot can simultaneously observe and feel the result of that contact. Different outcomes of this interaction help not only to determine the outline of the object against the background, but also physical characteristics of it. For example, heavy objects halt the arm's movement. Lighter objects may begin to move, and the resulting flow field indicates the boundary of the object. Complex objects may be more amenable to movement along different axes as a result of their geometry or friction properties, or both, which the robot can discover by poking an object from different directions. The robot thus demonstrates a form of active perception (Noë 2005), in which the object is understood not so much as a result of its external appearance, but rather by the affordances (Gibson 1977) that the object projects: in other words, the ways in which the embodied agent may interact with the object. Indeed active perception is a growing area of study in evolutionary robotics (Gomez and Eggenberger 2007; Tuci, Massera, and Nolfi 2009; Bongard 2009a, 2009b).

Metta and Fitzpatrick also point out that this blending of perceptual and motor processes is observed in the primate brain, and that this blending may therefore have a functional role rather than being an accident of evolution. Finally, they demonstrate that this interaction with the environment can support the development of more complex cognitive abilities such as mimicking a human demonstrator. The main long-term goal of artificial intelligence is, after all, the realization of more complex cognition through the aggregation of simpler competencies; this work, along with others (e.g., Yamashita and Tani 2008) indicates how embodied behavior may simplify this progression.

6.4   Morphology Increases Evolvability

The ability to find good solutions within a search space is dictated primarily by three factors: the dimensionality of the search space, its smoothness, and the degree of neutrality. Smoothness can be operationally defined as how much phenotypic effect a small genetic mutation causes. Mutational effect can be influenced by the amount of genetic interaction, as in Kauffman's NK fitness landscapes (Kauffman and Johnsen 1991), where N represents the dimensionality of the genotype and K indicates the degree of coupling between genes. For higher K values, a slight genetic mutation will influence the expression of several other genes, magnifying the mutation's phenotypic effect. It has been known for some time in biology that the probability of a mutation conferring a fitness benefit is inversely proportional to the phenotypic magnitude of that mutation (Fisher 1930). Therefore, for evolutionary algorithms with high epistasis, the resulting fitness landscape tends to be "rugged": regions of high fitness are separated from one another by difficult-to-cross regions of low fitness.

The effect of neutrality on evolvability[2] has also been extensively studied in biology (Kimura 1983) and evolutionary computation (Barnett 1998; Yu and Miller 2002), and shows that populations may discover more high fit regions by undergoing a series of neutral mutations. Neutral mutation in a population can be visualized within the fitness landscape metaphor by envisioning a series of solutions clustering upward along a slope of increasing fitness, but then diffusing outward across a horizontal plateau, and thereby discovering multiple routes upward toward more fit solutions.

It follows from this that increasing the dimensionality of the search space for a given problem does not necessarily make it more difficult for an evolutionary algorithm to discover good solutions, as long as the correspondingly larger search space is smoothed in the process. In other words, a larger yet smoother fitness landscape may be more amenable to evolutionary search than a smaller, more rugged landscape.

Adding additional dimensions to a search may indeed smooth it in the process by introducing what the theoretical biologist Michael Conrad termed "extradimensional bypasses" (Chen and Conrad 1994; Cariani 2002): isolated fitness peaks in the small space may become saddle points in the larger space. This additional genotypic material may not necessarily introduce phenotypic novelty: rather, it may introduce intermediate, highly-fit solutions between two solutions that already existed in the smaller space such that an evolving population more often finds the higher peak by traversing the extradimensional bypass.

Of course, blindly increasing the dimensionality of a search space does not guarantee the creation of bypasses: larger search spaces might just as easily become more rugged than smooth. Seen through the lens of evolutionary robotics, we may take an existing ER experiment in which only the controller is optimized, increase the genotype to also specify some morphological parameters, and then re-evolve robots against the same task. We may expect one of three outcomes. First, the new experimental regime may turn out to be less evolvable than the original experiment, suggesting that the increased search space has either kept the ruggedness of the landscape constant or increased it. Second, the evolutionary algorithm may outperform the original experiment because it discovers a different body plan than the fixed one used previously. Third, performance may increase but the final, best robot may have the same or very similar body plan to the fixed one from the original experiment.

This last outcome would suggest that the additional evolvable morphological parameters have introduced extradimensional bypasses into the search space: evolution gradually modified the initial, default body plan through a series of intermediate forms—as well as modifying the controllers accompanying those body plans—until eventually evolution converged back on a body plan similar to the original body plan.

This evolutionary trajectory has been observed in several independent studies that employed different evolutionary algorithms, robots, and tasks. In Bongard and Paul (2001) we first evolved a simulated biped with an active torso to walk over flat terrain (figure 6.1o while keeping the morphology fixed. A standard genetic algorithm was used to evolve a feedforward neural network to optimize the robot's displacement over a fixed time period. In this first set of experiments the radii of the lower and upper legs and arms were kept constant. In the second set of experiments these three morphological parameters (lower leg, upper leg, and arm radii) were evolved along with the robot's neural network controller. In some of these latter experiments evolution discovered body plans different from the default, such as the one shown in figure 6.1o. It can be seen that the radii of the legs and arms are significantly larger than that of the torso, producing a bimodal mass distribution: mass is gathered in the legs to stabilize walking; and torque induced by walking is canceled by the neural network swinging the heavy arms in opposition to the legs. Other experiments produced robots that outperformed those found in the original set of experiments, but had very similar morphologies: in these runs the body plan was observed to change, but converged eventually on the original, default morphology. From these observations we concluded that although the additional morphological parameters had increased the dimensionality of the search space, it had also created new fitness peaks (as illustrated by the robot in figure 6.1o) as well as introduced extradimensional bypasses that allowed evolution to more often climb to fitness peaks that existed in the original experiments but which were more difficult to find.

More recently, I evolved a simulated robot manipulator based on the human arm and hand to grasp, lift, and actively distinguish between several objects. Each experiment series in which only the controller of the arm was evolved was paired with a second experiment series in which the radii and length of each finger's phalanges and the distribution of fingers around the hand were evolved (figure 6.3).

In the fixed, default morphology the fingers are arranged equidistantly around the spherical palm, and the lengths and radii of the phalanges were set to the same value. When the morphology was evolved, initial robots had random controllers but started with the default morphology. Evolution could then place fingers closer or further from each other around the palm, and phalange lengths and radii could differentiate. It was

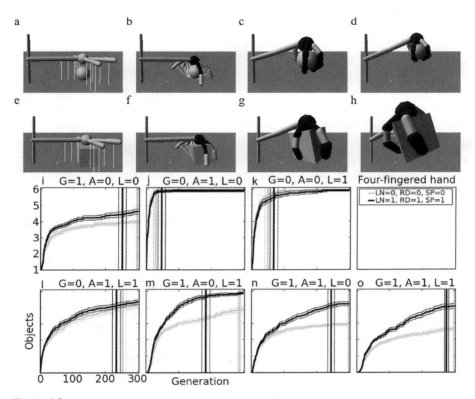

Figure 6.3
Evolving the morphology and controller of an anthropomorphic arm. (a–h) An evolved arm capable of grasping, lifting, and actively distinguishing between objects of different shapes. (a–d) The resultant behavior of the arm when grasping spheres. (e–h) The resultant behavior when the same arm grasps cubes. Note the trajectories of the arm in the two cases are different, indicating it can distinguish between them. The thin white lines indicate range sensors. Black and white finger segments indicate whether the tactile sensor in that segment is firing (black) or not (white). The evolved, differing radii of the phalanges can be seen most clear in (h). (i–o) Fitness improvements over evolutionary time for arms with fixed morphology (gray curves; 30 independent runs each) and evolved morphology (black curves; 30 independent runs each). In the former case phalange length (LN), phalange radii (RD), and spacing (SP) between phalanges was not evolved (LN,RD,SP = 0); in the latter case all three types of morphological parameters were evolved (LN,RD,SP = 1). Evolutionary time is measured in generations along the horizontal axis; the vertical axis indicates the mean number of generations required for the arms to successfully manipulate the object (two of the six objects are shown in [a–h]). The grey and black vertical lines indicate the mean time until the arms successfully evolved to manipulate all six objects. In different regimes, arms were evolved to grasp (G), actively distinguish between (A) and/or lift (L) the objects. Unity and zero indicate whether that aspect of manipulation was selected for or not, respectively. Thick lines indicate means; thin lines indicate one unit of standard error. For more details please refer to Bongard 2009b.

hoped that this would lead to the appearance of specialized digits such as human's opposable thumbs; indeed in several (but not all) runs differentiated digits were observed.

Again, I found that despite the expanded search space, due to the inclusion of the morphological parameters, evolution consistently found fitter behaviors compared to when hand morphology was held fixed. Also, in many runs the shape and distribution of the fingers would drift away from the default morphology and then gradually converge back to the default.

The creation of extradimensional bypasses through the evolution of morphology suggests the morphology should not only be optimized because a body plan different from that envisioned by human engineers may be found. Rather, it is often the *process* of gradually changing morphology that increases performance: there may be particular morphologies in which it is easier to discover a workable but crude controller, but that are not ideal for optimal performance. Subsequent evolution may refine the control and morphology (perhaps returning morphology close to the originally fixed body plan) so that robots may exhibit performance close to this optimal.

For example, it was observed in Bongard and Paul (2001) that early in some evolutionary runs mass was focused in the legs, making the morphology quite stable. This allowed for the discovery of a shuffling gait. Later evolution moved mass into the arms, thereby allowing for longer strides and using the heavy arms to cancel the torque induced by such a gait. This could be viewed as a kind of scaffolding (Wood, Bruner, and Ross 1976), as it is known in psychology, or shaping (Dorigo and Colombetti 1994), as it is known in the robotics literature: typically scaffolding or shaping implies that a learner's environment is structured to facilitate learning; here, the learner's own body provides a gradient for discovering crude behaviors and then refining them as the learner's body plan changes. I refer to this additional kind of scaffolding as *morphological scaffolding*.

6.5 Morphology Affords New Behaviors

In most robotics experiments, the robot's body plan is fixed by the experimenter. In several of the research projects surveyed so far in this chapter, body plans change over evolutionary timescales. In a particular class of robots—modular robots—there is the possibility of morphological change over the lifetime of the robot itself. A modular robot is typically made up of independent units, or modules that can attach to and detach from one another either under active control of the robot, or as response to environmental perturbation. This latter class of modular robots are known as stochastic modular robots (White et al. 2005); a simulated and physical example are shown in figure 6.1m,n. Behaviors exhibited by modular robots are typically denoted using a self-X nomenclature: such systems may start as independent modules and gradually attach into some desired form (self-assembly [Murata, Kurokawa, and Kokaji 1994; White et al. 2005; Groß et al. 2006]); create a copy of themselves (self-reproduction [Zykov et al. 2005]); start in some desired form, experience module loss or malfunction, and reconfigure back into the original form (self-repair [Tomita et al. 1999; Stoy and Nagpal 2004]); or alter form to exhibit different behaviors (self-reconfiguration [Murata et al. 2002; Park et al. 2008]).

This particular approach to embodied cognition creates a rich new vista of behaviors beyond the reach of traditional robots, or for that matter beyond most biological organisms.[3] Using the self-assembling machine and associated simulator shown in figure 6.1m,n, we explored different strategies for stochastic self-assembly (White et al. 2005). One of the main goals of modular robotics is to scale down the units to micro- or nanoscale. Such scales enforce strict constraints on the internal complexity of the units, so one approach is to assume units have no internal power source or actuation machinery: rather, the modules are moved by the flow of their surrounding motion. We further assumed that each side of each cube can be magnetized, and could attach to a powered floor plate if coming into contact with it. Once powered, an attached module can flip the polarity of the magnets on any of its exposed sides, thereby attracting unpowered modules suspended in the medium. If a module attaches to the structure it becomes powered and may in turn attract other modules, allowing for self-assembly.

By specifying a sequence of commands that flip magnet polarity during the self-assembly process, different structures can be created such as a vertical tower (figure 6.1m). However, for more complex structures attracting independent modules to particular surfaces may be difficult. For example if a cage-like structure is self-assembling it may be very improbable for independent modules to ricochet into the internal volume of the cage through random motion. However, by flipping the polarity between two attached modules, substructures can be jettisoned back into the fluid to be attached elsewhere. This allows for the added capability of self-reconfiguration, which can be exploited to overcome this limitation.

In White et al. (2005) we described two programs that specified the assembly of the same structure: in the first program modules were only attached; in the second program, modules were attached and detached during different stages of the assembly process. This latter program was more complex than the former, yet it led to more rapid mean assembly times. The reason for this was that attached modules were detached at strategic times and places. Once released back into the medium, they had low linear and rotational velocity, and were proximal to surfaces that would otherwise have had difficulty attracting an independent module. These surfaces would then often capture the recently released module, and the self-assembly process would continue.

This result demonstrates yet another advantage of building robot systems in which morphology may change: the robot's growing body may serve as its own scaffold, easing the otherwise difficult task of self-assembly. This is another example of the morphological scaffolding concept introduced in the previous section. When evolving robot body plans along with controllers, body plans change over evolutionary time and increase the probability of discovering a robot that can accomplish the desired task. In stochastic self-assembling modular robots, the robot can change the topology of its own body over the lifetime of the system such that it increases the probability of assembling the desired structure.

In another set of experiments from the same work, we investigated assembly times for structures composed of the same number of objects, but having different geometries: we studied structures ranging from a vertical tower (figure 6.1m) to branched and then closed structures. We found that despite the same volume of these structures, the closed structure was assembled more rapidly and consistently than the branched or tower structure. This indicates another importance of considering morphology for such systems: one should not only consider which body plan of a modular robot is appropriate for a given task once assembled, but also which body plan also has the best probability of self-assembling (or self-reconfiguring or self-repairing) in its environment.

Finally, Zykov et al. (2005) described a self-reproducing robot, in which the body plan of the parent robot played an important role in self-reproduction. The parent robot is composed of a series of magnetic modules, bonds to external modules, and deposits them on to the growing child robot, thereby dynamically changing the topology of its body in the process. The child robot's body plan also plays an important role in the self-reproduction process. It rotates itself to accept donated modules from the parent robot at the right place on its body, thereby contributing to a process in which the body plan of both parent and child robot change over time.

In the described experiments the self-assembly, self-reconfiguration, and self-reproduction programs were hand-designed. The finding that self-assembly can be deceptive (i.e., simpler programs do not always produce the desired structure more often or consistently than a more complex program) suggests that human intuition is difficult to apply to this problem, and that an automated process may therefore produce superior self-assembly programs. However, the number and type of commands as well as their timing are not known a priori, so a learning method in which a fixed number of parameters are optimized may not be appropriate. Rather, an evolutionary algorithm such as genetic programming (Koza 1992), in which the structure and parameters of algorithms can be optimized, may be appropriate in this domain in future. Indeed preliminary work (Estévez and Lipson 2007) has investigated evolving self-assembly programs in which the fitness of a program is the rapidity and consistency with which it leads to the desired structure.

6.6 Morphology Supports Self-Exploration

One of the necessary abilities an agent must possess to be considered intelligent is the capacity for learning. Learning has long played a role in AI and robotics, predating evolutionary robotics by several decades. Developmental robotics (Lungarella et al. 2003), another sister discipline of evolutionary robotics (in addition to biorobotics), is dedicated to investigating how a robot can exploit its body to scaffold its own learning, although typically the body does not change during this process. The experiments described in section 6.3 illustrate a typical experiment from that field.

Work within machine learning, a branch of disembodied AI, is often partitioned into passive learning, in which the learner passively receives the raw material for learning from a teacher, and active learning in which the learner must harvest its own material. Much work in that field has investigated strategies how to extract the most informative data for learning from the external world (Seung, Opper, and Sompolinsky 1992; Baram, Yaniv, and Luz 2004).

When the learner has a physical body, the challenge becomes how to best exploit its body to extract useful information about itself and its environment. This question of course depends on the goal of the learner. In past work (Bongard, Zykov, and Lipson 2006) we introduced a physical, autonomous robot that is able to autonomously generate legged gaits for itself and regenerate a new gait if it suffers unanticipated body plan change such as physical damage. In order to accomplish this, the robot evolves simulations that reflect the current topology of its body plan; if it is damaged, it re-evolves these simulations to reflect its changed state.

In most real-world contexts it is infeasible to equip a machine with enough sensors to detect all possible malfunctions. Therefore, it was assumed that the robot would use inferred relationships between motor commands and subsequent sensor signals to infer possible changes in its morphology. The robot therefore performs a short motor sequence (figure 6.4a), records the resulting sensor signature, and then generates a population of simulations, each of which contains a different body plan topology (figure 6.4b). The robot actuates each simulation with the same motor program it just performed, and compares the resulting sensor data from the simulation against the physical sensor data: the closer the match, the more accurate the simulator must be. Using an internal evolutionary algorithm the robot can then optimize the models to better reflect the topology of its body plan.

However, a single motor program is unlikely to provide sufficient information for it to accurately assess the state of all of its body parts. Therefore the robot must perform several actions to extract sufficient information for self-modeling. But we cannot allow the robot to perform an arbitrary number of actions, as physical motion is expensive in terms of time, power, and the risk of suffering additional damage by performing inappropriate actions in poorly understood environments. Therefore the robot must

Figure 6.4

An example of an autonomous robot that integrates self-modeling and internal behavior generation. The robot executes an action (a) and then creates a set of models to describe the sensorial result of that action (b). It then uses the models to find a new action (c) that will reduce uncertainty in the models when executed. By alternating between modeling and testing, it eventually finds and then uses an accurate model to internally optimize a behavior (d) before executing it in reality (e,f). (Courtesy of AAAS) employ an active learning methodology that finds motor programs likely to extract the most information from the external environment. If successful, the robot can minimize the number of actions it has to perform.

The robot uses query by committee (Seung, Opper, and Sompolinsky 1992) as the underlying active learning method. A population of initially random self-models is evolved against a single motor program and the resulting sensor data. Model evolution is paused after a short period and a second evolutionary algorithm searches for a motor program that, when supplied to the current model population, causes them to output diverging time series sensor data (figure 6.4c). This informative motor program is executed by the physical robot, the resulting sensor data is recorded, and the models are re-evolved to explain both motor programs. This process continues for a fixed number of cycles or until a sufficiently accurate self-model is found. An accurate self-model is then used in the traditional evolutionary robotics manner: a controller is optimized aboard the simulated robot such it performs the desired task, after which the best controller is executed by the physical robot (figure 6.4d).

In this domain in which the body plan of the physical robot may change unexpectedly,[4] morphology drives behavior at several levels: the overall learning framework uses the physical body as a vehicle for generating and extracting information from the world, and the simulated morphologies in the self-models influence which motor program will be executed. It was demonstrated in Bongard, Zykov, and Lipson 2006 that this method can be used to automatically create a simulator for a robot rather than hand-design one, and can allow the robot to continuously generate compensating behaviors in the face of unexpected damage, malfunction, or degradation.

This work also goes some way toward reconciling the seemingly divergent philosophies of traditional and embodied AI: while the former stressed purely computational notions such as modeling and planning, the latter demonstrated that robust behavior could be realized through model-free, sensor-to-motor coupling (Brooks 1991a, 1991b). Our robot exhibited several automatically generated gaits such as the one shown in figure 6.1e that result from tight sensor-motor coordination. However, these gaits are the result of a separate process that uses exploratory actions to create internal self-models, and self-models to drive exploratory actions. This suggests that the two branches of AI may not be as irreconcilable as was previously thought: it may be possible to gradually augment the cognitive capabilities of a robot by grounding them in low-level sensor-motor processes. Regardless, this reconciliation is only possible by placing the physical robot's body at the center of cognitive processes, rather than focusing on cognition while marginalizing or discarding the body.

In neuroscience, there is a similar reconciliation between abstract thought and sensor-motor coordination, which is often referred to as mental simulation or motor imagery (Porro et al. 1996; Hesslow 2002). Mental simulation implies that higher brains "run" candidate motor programs and reason about the results. This contrasts with older, more disembodied theories in which the brain extracts concepts from their sensorimotor components, stores them in mental symbols, and higher-thought processes such as planning and reasoning transform those symbols. The role of symbols in AI and cognitive science raises difficult issues such as how abstract symbols can acquire real-world meaning (Harnad 1990). These issues somewhat dissipate when a closer connection between sensor-motor processes and cognitive processes is forged, such as in the robot just described. The robot's self-models have real-world meaning insofar as they arise from sensor-motor processes, and they allow the robot to determine how it may move. In this sense self-models *afford* (Gibson 1977) possible action: they tell the robot how it may interact with the world.

Although it is not yet known how biological brains simulate action, there is growing evidence that the mental toolkit for this exist. Body images encoded as topographic maps exist in the primate brain, and often exhibit topologies similar to the sensor and motor systems that project to them: the retina projects to the primary visual cortex; the organ of Corti to the primary auditory cortex; and cutaneous receptors to the primary somatosensory cortex (Kandel, Schwartz, and Jessell 2000). Also, both forward and inverse models are found in the cerebellum (Wolpert, Miall, and Kawato 1998; Wolpert and Kawato 1998): like the robot just described, the primate brain can predict future sensor state given a candidate motor primitive by supplying that primitive to a forward model, and inverse models can be used to find an appropriate motor primitive that will produce a desired future sensor state. Although it is unknown how higher brain areas like the cerebral cortex make use of these sensorimotor maps and models, it seems likely that there is no clear dividing line in the brain between sensorimotor representation and simulation and higher cognition.

6.7 Morphology Creates New Research Questions

As just described, how situated action supports the emergence of more complex (and possibly cognitive[5]) behaviors is becoming a hot topic in evolutionary robotics and AI in general. However, fixing a robot's morphology places limits and biases on the kinds of action that the robot can perform, and therefore also on the more complex behaviors that those actions may eventually support. A robot with legs can only exhibit legged locomotion; a wheeled robot with a rigid gripper can only move over flat terrain and grasp objects with a fixed radius. Therefore, given a desired task, a roboticist should select not only appropriate control architecture for the robot, but also an appropriate body plan.

The engineering tradition places an inordinate emphasis on modularity (Suh 1990): presumably, this is primarily due to the reductionist paradigm that emphasizes breaking problems down into separable subproblems, each of which can be more easily solved than the original problem. The general concept of modularity can be broken down into functional and structural modularity: what are the separate functions that an agent (or its constitutive parts) must perform, and what are the substructures that make up an agent's phenotype? Broadly speaking, structure dictates function: the shape of a protein completely determines its function; the morphology of the primate hand determines its possible grasping strategies. However, it does not follow that a structural module must contribute to only one function: in addition to grasping, the primate hand contributes to a near infinitude of other functions from fine manipulation to whole-body brachiation.

Yet in robotics one tends to observe a one-to-one correspondence between morphological components and function: in a wheeled robot with a gripper, wheels contribute to movement but not grasping, while a gripper contributes to grasping but not locomotion. This seems on the surface to simplify things: the researcher can design a subcontroller for each function that interacts only with the substructure responsible for that function. In this spirit several projects have explicitly evolved modular neural networks in which each neural module contributes to a different function (Brooks 1986; Calabretta et al. 2000). However, this approach is not scalable: humans are capable of a very large number of behaviors, and it is unlikely that there are separate brain structures for each of them. What seems more likely is that similar actions are driven by common neural circuitry. Indeed in the field of evolutionary robotics several researchers have demonstrated that this explicit structural modularity in the robot's neural network is not always necessary (Bongard 2008; Izquierdo and Buhrmann 2008; Auerbach and Bongard 2009b).

In recent work (Auerbach and Bongard 2009a) we have explored a simulated robot morphology (figure 6.5) that challenges the structural and functional modularity observed in other morphologies. The robot is composed of a series of segments, each of which supports a pair of cylindrical appendages. Each appendage pair can be actuated, as can intersegmental joints that allow the robot to flex its "spine" within the sagittal plane. This allows the robot to rotate its upper or lower body up- and downward, or to keep its body horizontal during locomotion. In this work the robot's morphology was not evolved; only its controller was. The fitness function selected for robots that locomote toward an object in their environment, grasp the object once reached, and lift it. The robot is equipped with sensors that signal the distance of the sensor from the object at each time step of the simulation: the sensors can be thought of as returning a signal commensurate with ambient sound amplitude, and the object as emitting a continuous sound. The robot's front appendages were also equipped with touch sensors.

Given the body plan and task environment of the robot, there are two classes of behaviors that the robot may adopt to complete the task. In the first class, the robot would raise the front of the body, locomote to the object using the back four appendages, and manipulate the object with the front pair when the object is reached. The

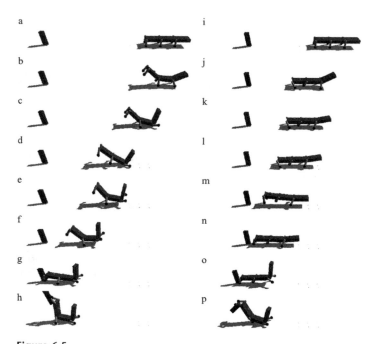

Figure 6.5
Two evolved behaviors for a segmented robot. (a–h) A sample functionally specialized controller. The robot initially raises its front pair of appendages and keeps them raised while it approaches the target object. Once reached, it grasps and lifts the object. (i–p) A functionally generalized controller. The front appendage pair participates in locomotion toward the object. Once reached, the front pair participates in object manipulation.

second strategy involves developing a hexapodal gait such that all six appendages are involved in locomotion toward the object; once reached, the front of the body would rise to allow the front appendage pair to manipulate the object. Both such strategies, if performed rapidly, achieve a very high fitness value. It was found that in different independent trials both strategies were discovered by evolution: figure 6.5a shows an example of the first class of behaviors; figure 6.5b shows an example of the second class.

It was also found that functionally specialized controllers tended to arise much more often than functionally generalized ones, even though both, when optimized, obtain about the same fitness values. This raises the question as to why and how specialization arises. We investigated three hypotheses: (1) functionally specialized controllers are more evolvable (i.e., the slopes of fitness peaks surrounding specialized controllers can be more easily climbed than peaks surrounding generalized controllers); (2) early discovery of rudimentary specialized controllers is more common, and the degree of specialization is difficult for evolution to tune throughout a run so evolution is forced to refine these behaviors; or (3) functionally specialized controllers more easily allow for active perception (Noë 2005).

The first hypothesis seems unlikely due to the observation that when comparing runs that discover specialized controllers and those that discover functionally specialized controllers, both exhibit relatively rapid, smooth, and consistent fitness improvement over the run. This suggests that both kinds of solutions afford relatively smooth gradients for the population to climb. Many runs exhibited controller phylogenies that gradually changed from specialized to generalized and back again, providing evidence against the second hypothesis.

This leaves the third hypothesis, which seems to be supported by the results. When the front of the body is raised early during an evaluation, the touch sensors fall silent. This silence can be shaped by evolution into a trigger that guides the controller into a cyclic attractor, which produces a rhythmic gait toward the object. If the front appendages come into contact with the object then the touch sensors fire: again, evolution can alter the controller such that this signal shifts the controller into a different dynamic state that causes object grasping and lifting. This is a form of active perception in that through the robot's interaction with the environment—mediated by its morphology—the robot either approaches or manipulates the object.

By choosing this segmented body plan, new research questions arise, namely, how will the responsibility for different behaviors be distributed across the agent's controller *and* morphology? And how will evolution accomplish this, given the opportunity to do so? If we wish to realize robots capable of ever more complex tasks, these robots will have to perform several behaviors separated in time, space, and across their morphological and control substructures. It seems unlikely that each such substructure will support one and only one function, but how to overlap them is nonintuitive. Therefore it will become important to investigate these new research questions by considering different combinations of robot morphology and control. More desirable, however, is to allow evolution to co-optimize both to discover the most appropriate distribution of function across the robot's structure.

6.8   Morphology Supports Scalability

The preceding six lines of argument are unlikely to convince many in the robotics community that formulating an appropriate robot morphology for a given task is nonintuitive. In order to make progress along this front it is necessary to clearly demonstrate under what conditions this intuition breaks down. For this reason I conducted a series of experiments with an anthropomorphic arm (already briefly described in section 6.4 and illustrated in figure 6.3a–h) in which both the morphology and the controller of the arm were evolved. The arm was exposed to a variety of selection pressures that favored the manipulation of a series of objects in some way. Three fitness functions were formulated that selected for object grasping, lifting, and active categorical perception. Grasping (G) is accomplished by closing the fingers tightly around objects, and lifting (L) is accomplished by maintaining contact with the object while raising the arm. Active categorical perception (A) is accomplished by causing the controller to exhibit common signal signatures among the hidden neurons when grasping objects of the same shape but different signatures when manipulating objects of different shape.

Different selection regimes could be created by evolving a population of robots using multiobjective optimization (Deb 2002), in which each of these competencies is an objective. This allowed for a total of seven regimes in which robots were evolved against only one objective (G = 1, A = 1, or L = 1; figure 6.3i–k), any two objectives (G + A + L = 2; figure 6.3l–n), or all three objectives (G,A,L = 1; figure 6.3o). Shaping (Dorigo and Colombetti 1994) was also employed in that, at the outset of an evolutionary run, robots were only evolved against a single object. When fitness rose to a prespecified threshold a second object was added to the training pool, and so on up to a total of six objects.

It was found, not surprisingly, that the more objectives a robot population had to satisfy simultaneously, the more difficult it was to evolve a robot that could successfully manipulate all six objects (note the lowered curves in figures 6.3l–o compared to figures 6.3j–k[6]). In addition, it was found that evolving aspects of the hand's morphology along with the controller, compared to just evolving the hand's morphology, did not provide any advantage when only one objective was selected for.[7] However, for two or three objectives, evolving morphology along with control provided a significant performance advantage.

This result suggests that at least for object manipulation, an intuitive biologically inspired hand design—in which four fingers with more or less equal phalange length, radii, and equal separation between the digits—is perfectly serviceable if the task is relatively simple. However, when the robot must satisfy several goals simultaneously (such as grasping, lifting, and actively distinguishing between objects of different shape), slight changes in this default morphology provide dramatically increased capability. For example, in one independent trial in which all three manipulation capabilities were selected for, the robot shown in figure 6.3a–h was evolved. Note the slight differences in phalange radii within individual digits. This ensures that when the hand closes on an object, different phalanges come in contact with the object depending on the object's shape. This in turn causes different touch sensors to fire for differently shaped objects, stimulating different hidden neuron signatures and satisfying the active categorical perception requirement. Such combinations of a tight grip and active perception are more difficult if phalanges on the same digit have the same radii.

This provides evidence that as roboticists we may indeed be able to formulate perfectly serviceable morphologies for simple tasks, but as the tasks become more difficult, for example by requiring that the robot accomplish several tasks at once, our intuitions as designers break down. Even slight changes to the body plan may greatly simplify the task (as discussed in section 6.3 or allow for simpler controller policies (section 6.2, although which changes provide these advantages may not be immediately clear: they may however be found through evolutionary search.

6.9   Conclusions

This chapter has surveyed the history and current state of an intellectual movement in artificial intelligence centered on the concept of embodiment: in order to exhibit intelligent behavior, the agent must have a body with which to interact with the environment. This movement can be summarized by considering the progression of research questions that have been asked: *what*, *how*, and *why*.

The founders of embodied AI (Brooks 1991a; Clark 1996; Pfeifer 1999; Pfeifer and Bongard 2006) described *what* embodiment is, and what some of the consequences of having a body are. This raised the issue of *how* to choose an appropriate body plan for a given task. Evolutionary robotics provided one answer to this question by showing how evolutionary algorithms could be extended to optimize both the morphology and control policies of autonomous robots (Prusinkiewicz and Lindenmayer 1990; Sims 1994; Ventrella 1994; Eggenberger 1997; Lipson and Pollack 2000; Hornby and Pollack 2001; Bongard 2002). If evolutionary robotics is to be considered a serious engineering or computational discipline, or both, however, we must provide concrete evidence of increased performance to justify the cost of this additional algorithmic complexity. In other words, we must provide reasons for *why* evolutionary algorithms are employed for morphology optimization.

This chapter has described seven such reasons. A robot body plan appropriate to the task at hand can simplify control and make seemingly difficult tasks easier, but it is often difficult to determine manually what morphology would provide these benefits. Allowing evolution to optimize morphology instead requires additional evolvable parameters and therefore increases the dimensionality of the search space, but if done properly this can actually smooth the enlarged search space and increase the probability of finding good solutions. The burgeoning field of modular robotics assumes that the topology of the robot's body plan can change over its lifetime, which generates new behaviors beyond traditional robotics such as self-assembly, self-repair, self-reconfiguration, and even self-reproduction at several size scales. A curious robot with the right morphology can learn intelligently about its own capabilities and limitations, as well as discover unforeseen changes to its body as a result of damage. The adoption of nonstandard morphologies such as segmented body plans raises new research questions about how to distribute the responsibility for several behaviors across different parts of the robot body. Finally, as we attempt to evolve robots capable of ever more complex behaviors, our intuitions about what form the robot's body must take break down, and we must rely on artificial evolution to explore the rich space of physical forms that support such behaviors—just as biological evolution did.

Notes

1. It is often difficult to decide which aspects of biological body plans to mimic.

2. Evolvability can be loosely defined as the ability of an evolutionary system to consistently discover higher-fit phenotypes.

3. The notable exception is certain slime molds that transition between single- and multicellular configurations.

4. Rather than deliberately, as in the modular robotics examples.

5. "Cognition" is a controversial term: here I use it as shorthand for sufficiently complex behaviors that may cause an outside observer to consider the robot to be, to some degree, cognitive. Of course, for different observers this threshold will be different, rendering "cognition" a subjective term.

6. The exception to this was grasping (figure 6.3i), which benefited when lifting or active perception, or both, was also selected for.

7. Again, with the exception of grasping, in which evolving morphology provided a slight benefit (figure 6.3i).

References

Auerbach, J., and J. C. Bongard. 2009a. Evolution of functional specialization in a morphologically homogeneous robot. In *Proceedings of the Genetic and Evolutionary Computation Conference*, 89–96. Montreal, Canada.

Auerbach, J., and J. C. Bongard. 2009b. How robot morphology and training order affect the learning of multiple behaviors. In *Proceedings of the IEEE Congress on Evolutionary Computation*, 39–46. Trondeim, Norway.

Baram, Y., R. E. Yaniv, and K. Luz. 2004. Online choice of active learning algorithms. *Journal of Machine Learning Research* 5:255–291.

Barnett, L. 1998. Ruggedness and neutrality: The NKp family of fitness landscapes. In *Proceedings of the Sixth International Conference on Artificial Life*, ed. C. Adami et al., 18–27.

Bechtel, W. 1990. Connectionism and the philosophy of mind. In *Mind and Cognition: A Reader*, ed. W. G. Lycan, 252–273. Oxford: Basil Blackwell.

Bongard, J. C. 2002. Evolving modular genetic regulatory networks. In *Proceedings of the 2002 Congress on Evolutionary Computation*, 17–21.

Bongard, J. C. 2008. Behavior chaining: Incremental behavioral integration for evolutionary robotics. In *Proceedings of the Eleventh International Conference on the Simulation and Synthesis of Living Systems*, ed. S. Bullock et al., 64–71.

Bongard, J. C. 2009a. Accelerating self-modeling in cooperative robot teams. *IEEE Transactions on Evolutionary Computation* 13 (2): 321–332.

Bongard, J. C. 2009b. The utility of evolving robot morphology increases with task complexity for object manipulation. *Artificial Life* 16 (3): 201–223. doi: 10.1162/artl.2010.

Bongard, J. C., and C. Paul. 2001. Making evolution an offer it can't refuse: Morphology and the extradimensional bypass. In *Proceedings of the Sixth European Conference on Artificial Life*, ed. J. Keleman and P. Sosik, 401–412. Prague.

Bongard, J. C., and R. Pfeifer. 2001. Repeated structure and dissociation of genotypic and phenotypic complexity in artificial ontogeny. In *Proceedings of the Genetic and Evolutionary Computation Conference*, ed. L. Spector et al., 829–836.

Bongard, J. C., and R. Pfeifer. 2002. A method for isolating morphological effects on evolved behaviour. In *Proceedings of the Seventh International Conference on the Simulation of Adaptive Behaviour*, ed. B. Hallam et al., 305–311. Cambridge, MA: MIT Press.

Bongard, J. C., V. Zykov, and H. Lipson. 2006. Resilient machines through continuous self-modeling. *Science* 314:1118–1121.

Brooks, R. 1986. A robust layered control system for a mobile robot. *IEEE Journal on Robotics and Automation* 2 (1): 14–23.

Brooks, R. A. 1991a. Intelligence without reason. *Artificial Intelligence: Critical Concepts* 3: 569–595.

Brooks, R. A. 1991b. Intelligence without representation. *Artificial Intelligence* 47:139–159.

Brooks, R. A., C. Breazeal, M. Marjanovic, B. Scassellati, and M. M. Williamson. 1999. The Cog project: Building a humanoid robot. *Lecture Notes in Computer Science* 1562:52–87.

Calabretta, R., S. Nolfi, D. Parisi, and G. P. Wagner. 2000. Duplication of modules facilitates the evolution of functional specialization. *Artificial Life* 6 (1): 69–84.

Cariani, P. A. 2002. Extradimensional bypass. *Bio Systems* 64 (1–3): 47–53.

Chen, J. C., and M. Conrad. 1994. A multilevel neuromolecular architecture that uses the extradimensional bypass principle to facilitate evolutionary learning. *Physica D. Nonlinear Phenomena* 75:417–437.

Clark, A. 1996. *Being There*. Cambridge, MA: MIT Press.

Collins, S., A. Ruina, R. Tedrake, and M. Wisse. 2005. Efficient bipedal robots based on passive-dynamic walkers. *Science* 307 (5712): 1082–1085.

Deb, K. 2002. *Multi-Objective Optimization Using Evolutionary Algorithms*. New York: Wiley.

Dorigo, M., and M. Colombetti. 1994. Robot shaping: Developing situated agents through learning. *Artificial Intelligence* 70 (2): 321–370.

Eggenberger, P. 1997. Evolving morphologies of simulated 3d organisms based on differential gene expression. In *Proceedings of the Fourth European Conference on Artificial Life*, 205–213.

Estévez, N., and H. Lipson. 2007. Dynamical blueprints: Exploiting levels of system-environment interaction. In *Proceedings of the Genetic and Evolutionary Computation Conference*, 238–244.

Fisher, R. A. 1930. *The Genetical Theory of Natural Selection*. Oxford: Clarendon Press.

Frutiger, D. R., J. C. Bongard, and F. Iida. 2002. Iterative product engineering: Evolutionary robot design. In *Proceedings of the Fifth International Conference on Climbing and Walking Robots*, ed. P. Bidaud and F. B. Amar, 619–629. London: Professional Engineering Publishing.

Gibson, J. J. 1977. The theory of affordances. In *Perceiving, Acting and Knowing: Toward an Ecological Psychology*, ed. R. Shaw and J. Bransford, 67–82.

Gomez, G., and P. Eggenberger. 2007. Evolutionary synthesis of grasping through self-exploratory movements of a robotic hand. In *Proceedings of the IEEE Congress on Evolutionary Computation*, ed. A. Tay, 3418–3425.

Groß, R., M. Bonani, F. Mondada, and M. Dorigo. 2006. Autonomous self-assembly in swarmbots. *IEEE Transactions on Robotics* 22 (6): 1115–1130.

Harnad, S. 1990. The symbol grounding problem. *Physica D. Nonlinear Phenomena* 42:335–346.

Hesslow, G. 2002. Conscious thought as simulation of behavior and perception. *Trends in Cognitive Sciences* 6:242–247.

Hornby, G. S., and J. B. Pollack. 2001. Body-brain coevolution using L-systems as a generative encoding. In *Proceedings of the Genetic and Evolutionary Computation Conference*, 868–875.

Hunt, K. D. 1996. The postural feeding hypothesis: An ecological model for the evolution of bipedalism. *South African Journal of Science* 92 (2): 77–90.

Izquierdo, E., and T. Buhrmann. 2008. Analysis of a dynamical recurrent neural network evolved for two qualitatively different tasks: Walking and chemotaxis. In *Proceedings of the Eleventh International Conference on the Simulation and Synthesis of Living Systems*, ed. S. Bullock et al., 257–264.

Jablonski, N. G., and G. Chaplin. 1993. Origin of habitual terrestrial bipedalism in the ancestor of the hominidae. *Journal of Human Evolution* 24:259–280

Kandel, E. R., J. H. Schwartz, and T. M. Jessell. 2000. *Principles of Neural Science*. 4th ed. New York: McGraw-Hill.

Kauffman, S. A., and S. Johnsen. 1991. Coevolution to the edge of chaos: Coupled fitness landscapes, poised states, and coevolutionary avalanches. *Journal of Theoretical Biology* 149 (4): 467–505.

Kimura, M. 1983. *The Neutral Theory of Molecular Evolution*. Cambridge: Cambridge University Press.

Koza, J. R. 1992. *Genetic Programming: On the Programming of Computers by Means of Natural Selection*. Cambridge, MA: MIT Press.

Lichtensteiger, L., and P. Eggenberger. 1999. Evolving the morphology of a compound eye on a robot. In *Proceedings of the Third Workshop on Advanced Mobile Robots*, 127–134.

Lipson, H., and J. B. Pollack. 2000. Automatic design and manufacture of robotic lifeforms. *Nature* 406 (6799): 974–978.

Lovejoy, C. O. 1980. Hominid origins: The role of bipedalism. *American Journal of Physical Anthropology* 52:250.

Lungarella, M., G. Metta, R. Pfeifer, and G. Sandini. 2003. Developmental robotics: A survey. *Connection Science* 15 (4): 151–190.

McGeer, T. 1990. Passive dynamic walking. *International Journal of Robotics Research* 9 (2): 62–82.

Metta, G., and P. Fitzpatrick. 2003. Better vision through manipulation. *Adaptive Behavior* 11 (2): 109–128.

Minsky, M. 1974. *A Framework for Representing Knowledge*. Cambridge, MA: MIT Press.

Morgan, E. 1982. *The Aquatic Ape: A Theory of Human Evolution*. Souvenir Press.

Murata, S., H. Kurokawa, and S. Kokaji. 1994. Self-assembling machine. In *Proceedings of the IEEE International Conference on Robotics and Automation*, 441–448.

Murata, S., E. Yoshida, A. Kamimura, H. Kurokawa, K. Tomita, and S. Kokaji. 2002. M-TRAN: Self-reconfigurable modular robotic system. *IEEE/ASME Transactions on Mechatronics* 7 (4): 431–441.

Nelson, A. L., G. J. Barlow, and L. Doitsidis. 2009. Fitness functions in evolutionary robotics: A survey and analysis. *Robotics and Autonomous Systems* 57 (4): 345–370.

Noë, A. 2005. *Action in Perception*. Cambridge, MA: MIT Press.

Park, M., S. Chitta, A. Teichman, and M. Yim. 2008. Automatic configuration methods in modular robots. *International Journal of Robotics Research* 27 (3–4): 403–421.

Pfeifer, R. 1999. *Understanding Intelligence*. Cambridge, MA: MIT Press.

Pfeifer, R., and J. Bongard. 2006. *How the Body Shapes the Way We Think: A New View of Intelligence*. Cambridge, MA: MIT Press.

Porro, C. A., M. P. Francescato, V. Cettolo, M. E. Diamond, P. Baraldi, C. Zuiani, M. Bazzocchi, and P. E. di Prampero. 1996. Primary motor and sensory cortex activation during motor performance and motor imagery: A functional magnetic resonance imaging study. *Journal of Neuroscience* 16 (23): 7688–7698.

Prusinkiewicz, P., and A. Lindenmayer. 1990. *The Algorithmic Beauty of Plants*. New York: Springer Verlag.

Seung, H. S., M. Opper, and H. Sompolinsky. 1992. Query by committee. In *Proceedings of the Fifth Annual Workshop on Computational Learning Theory*, 287–294.

Sims, K. 1994. Evolving virtual creatures. In *Proceedings of SIGGRAPH*, 15–22. Orlando, FL.

Stoy, K., and R. Nagpal. 2004. Self-repair through scale independent self-reconfiguration. In *Proceedings of the IEEE/RSJ International Conference on Robotics and Systems*, 2062–2067.

Suh, N. P. 1990. *The Principles of Design*. Oxford: Oxford University Press.

Tomita, K., S. Murata, H. Kurokawa, E. Yoshida, and S. Kokaji. 1999. Self-assembly and self-repair method for a distributed mechanical system. *IEEE Transactions on Robotics and Automation* 15 (6): 1035–1045.

Tuci, E., G. Massera, and S. Nolfi. 2009. Active categorical perception in an evolved anthropomorphic robotic arm. In *Proceedings of the IEEE Congress on Evolutionary Computation*, ed. A. Tyrell et al., 31–38.

Ventrella, J. 1994. Explorations in the emergence of morphology and locomotion behavior in animated characters. In *Artificial Life IV: Proceedings of the Fourth International Workshop on the Synthesis and Simulation of Living Systems*, 436–441. Cambridge, MA: MIT Press.

Webb, B., and T. R. Consi. 2001. *Biorobotics: Methods and Applications*. Cambridge, MA: MIT Press.

White, P., V. Zykov, J. Bongard, and H. Lipson. 2005. Three-dimensional stochastic reconfiguration of modular robots. In *Proceedings of Robotics: Science and Systems*, 161–168. Cambridge, MA: MIT Press.

Wilson, F. R. 1998. *The Hand*. New York: Random House.

Wolpert, D. M., and M. Kawato. 1998. Multiple paired forward and inverse models for motor control. *Neural Networks* 11 (7–8): 1317–1329.

Wolpert, D. M., R. C. Miall, and M. Kawato. 1998. Internal models in the cerebellum. *Trends in Cognitive Sciences* 2 (9): 338–347.

Wood, D., J. S. Bruner, and G. Ross. 1976. The role of tutoring in problem solving. *Journal of Child Psychology and Psychiatry, and Allied Disciplines* 17 (2): 89–100.

Yamashita, Y., and J. Tani. 2008. Emergence of functional hierarchy in a multiple timescale neural network model: A humanoid robot experiment. *PLoS Computational Biology* 4 (11): e1000220.

Yu, T., and J. Miller. 2002. Finding needles in haystacks is not hard with neutrality. Lecture Notes in Computer Science. Berlin: Springer.

Zykov, V., J. C. Bongard, and H. Lipson. 2004. Evolving dynamic gaits on a physical robot. In *Late Breaking Papers for the 2004 Genetic and Evolutionary Computation Conference*, Seattle, WA.

Zykov, V., E. Mytilinaios, B. Adams, and H. Lipson. 2005. Self-reproducing machines. *Nature* 435 (7039): 163–164.

7 Evolutionary Swarm Robotics: A Theoretical and Methodological Itinerary from Individual Neurocontrollers to Collective Behaviors

Vito Trianni, Elio Tuci, Christos Ampatzis, and Marco Dorigo

7.1 Introduction

In the last decade, swarm robotics gathered much attention in the research community. By drawing inspiration from social insects and other self-organizing systems, it focuses on large robot groups featuring distributed control, adaptation, high robustness, and flexibility. Various reasons lay behind this interest in similar multi-robot systems. Above all, inspiration comes from the observation of social activities, which are based on concepts like division of labor, cooperation, and communication. If societies are organized in such a way in order to be more efficient, then robotic groups also could benefit from similar paradigms.

As Kube and Zhang (1993) have pointed out, "Constructing tools from a collection of individuals is not a novel endeavor for humankind. A chain is a collection of links, a rake a collection of tines, and a broom a collection of bristles. Sweeping the sidewalk would certainly be difficult with a single or even a few bristles. Thus there must exist tasks that are easier to accomplish using a collection of robots, rather than just one."

A multi-robot approach can have many advantages over a single-robot system. First, a monolithic robot able to accomplish various tasks in varying environmental conditions is difficult to design. Moreover, the single-robot approach suffers from the problem that even small failures of the robotic unit may prevent the accomplishment of the whole task. On the contrary, a multi-robot approach can benefit from the parallelism of operation to be more efficient, from the versatility of its multiple, possibly heterogeneous units, and from the inherent redundancy in using multiple agents (Jones and Matarić 2006).

Swarm robotics pushes the cooperative approach to its extreme. It represents a theoretical and methodological approach to the design of "intelligent" multi-robot systems inspired by the efficiency and robustness observed in social insects in performing collective tasks (Bonabeau, Dorigo, and Theraulaz 1999). Collective motion in fish, birds, and mammals, as well as collective decisions, synchronization, and social differentiation are examples of collective responses observed in natural swarms (for some recent reviews, see Camazine et al. 2001; Franks et al. 2002; Couzin and Krause 2003; Sumpter 2006; Couzin 2007).

In all these examples, the individual behavior is relatively simple, but the global system behavior presents complex features that result from the multiple interactions of the system components. Similarly, in a swarm robotics system, the complexity of the group behavior should not reside in the individual controller, but in the interactions among the individuals. Thus, the main challenge in designing a swarm robotics system is represented by the need to identify suitable interaction rules among the individual robots. In other words, the challenge is designing the individual control rules that can lead to the desired global behavior.

In the preceding perspective, self-organization is the mechanism that can explain how complex collective behaviors can be obtained in a swarm robotics system from simple individual rules. In this context, a complex collective behavior should be intended as some spatiotemporal organization in a system that is brought forth through the interactions among the system components. Not every collective behavior is self-organized, though (Camazine et al. 2001). The presence of a leader in the group, the presence of blueprints or recipes to be followed by the individual system components clashes with the concept of self-organization, at least at the level of description in which leader or blueprints are involved. Another condition in which a collective behavior cannot be considered self-organizing is when environmental cues or heterogeneities are exploited to support the group organization. For instance, animals that aggregate in a warm part of the environment following a temperature gradient do not self-organize. But animals that aggregate to stay warm, and therefore create and support a temperature gradient in the environment, do self-organize. In both cases, the observer may recognize the presence of some structure (the aggregate) that correlates with the presence of an environmental heterogeneity (the temperature gradient). However, the two examples are radically different from the organizational point of view. Similar natural examples can be easily given also for the presence of leader or blueprints, to show that not every collective behavior is self-organizing (Camazine et al. 2001). Both the leader and the blueprint can be recognized as the place where the behavioral complexity of the group is centralized. In other words, the complexity of the group behavior does not result from the multiple interactions among the individual behaviors. Rather, the group behavior results from a fixed pattern of interactions among the system components that is either decided beforehand (in the case of a blueprint) or is centrally or continuously re-planned, or both (in the case of a leader). In both cases, there is limited room for adaptiveness to unknown, unpredictable situations resulting from a highly dynamical environment, both physical and social.

The unpredictable nature of the (social) environment makes it difficult to predict in advance, and therefore design, the behavioral sequence and the pattern of interactions that would lead to a certain group behavior. Moreover, "the adaptiveness of an autonomous multi-robot system is reduced if the circumstances an agent should take into account to make a decision concerning individual or collective behaviour are defined by a set of a priori assumptions" (Tuci et al. 2006b). This design problem can be bypassed by relying on evolutionary robotics (ER) techniques as an automatic methodology to synthesize the swarm behavior (Trianni, Nolfi, and Dorigo 2008). In past researches conducted within the SWARM-BOTS project, we experimented with different tasks and defined a methodology that proved viable for the synthesis of self-organizing systems.

We focused on two particular kinds of self-organizing systems: (1) systems that are able to achieve and maintain a certain organization, and (2) systems close to a bifurcation point, where robot-robot interactions and randomness lead to one or the other solution. In both cases, the problem is solved without placing any assumption on the kind of interaction pattern that would have been exploited to achieve a certain goal. Even more important, we have shown that determining a priori a certain form of interaction may result in worse performance with respect to an assumption-free setup.

We present the SWARM-BOTS project's experience in section 7.2, and in section 7.3 we discuss in detail some examples of problems studied exploiting the ER approach. Then, in section 7.4 we speculate on the current limitations of the ER approach, and the future role of ER in the development of more complex behaviors and cognitive abilities for robotic swarms.

7.2   Swarm Robotics and the Swarm-bots

Even though research in swarm robotics is relatively novel, it is quickly developing thanks to the contribution of various pioneer studies (Kube and Zhang 1993; Beckers, Holland, and Deneubourg 1994; Holland and Melhuish 1999; Martinoli, Ijspeert, and Mondada 1999; Krieger, Billeter, and Keller 2000). The SWARM-BOTS project made a significant contribution to the field in the design and development of an innovative swarm robotics platform: the swarm-bot (Mondada, Floreano, and Gambardella 2004; Dorigo et al. 2004). A swarm-bot is defined as a self-assembling, self-organizing artifact formed by a number of independent robotic units, called s-bots. In the swarm-bot form, the s-bots become a single robotic system that can move and reconfigure. Physical connections between s-bots are essential for solving many collective tasks, such as the retrieval of a heavy object. Also, during navigation on rough terrain, physical links can serve as support when the swarm-bot has to pass over a hole wider than a single s-bot, or when it has to pass through a steep concave region.

However, for tasks such as searching for a goal location or tracing an optimal path to a goal, a swarm of s-bots can be more efficient. An s-bot is a small mobile autonomous robot with self-assembling capabilities, shown in figure 7.1. It weighs 700 g and

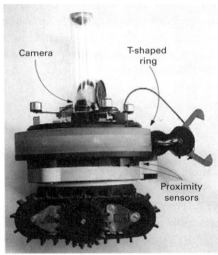

Figure 7.1
View of the s-bot from different sides. The main components are indicated (see text for more details).

its main body has a diameter of about 12 cm. Its design is innovative concerning both sensors and actuators. The traction system is composed of both tracks and wheels—referred to as "treels"—that provide the s-bot with a differential drive motion. The wheels are connected to the chassis, which contains the batteries, some sensors, and the corresponding electronics. The main body is a cylindrical turret mounted on the chassis by means of a motorized joint that allows the relative rotation of the two parts. The gripper is mounted on the turret and can be used for connecting rigidly to other s-bots or to some objects. The shape of the gripper closely matches the T-shaped ring placed around the s-bot's turret, so that a firm connection can be established. The gripper not only opens and closes, but also has a degree of freedom for lifting the grasped objects. The corresponding motor is powerful enough to lift another s-bot.

An s-bot is provided with many sensory systems, useful for the perception of the surrounding environment or for proprioception. Infrared proximity sensors are distributed around the rotating turret. Four proximity sensors placed under the chassis—referred to as "ground sensors"—can be used for perceiving holes or the terrain's roughness (see figure 7.1). Additionally, an s-bot is provided with eight light sensors uniformly distributed around the turret, two temperature/humidity sensors, a three-axis accelerometer and incremental encoders on each degree of freedom. Each robot is also equipped with sensors and devices to detect and communicate with other s-bots, such as an omni-directional camera, colored LEDs around the s-bots' turret, microphones, and loudspeakers (see figure 7.1). Eight groups of three colored LEDs each—red, green, and blue—are mounted around the turret. They can be used to emit a color that can represent a particular internal state of the robot.

The color emitted by a robot can be detected by other s-bots using the omni-directional camera, which allows the robot to grab panoramic views of the scene surrounding an s-bot. The loudspeaker can be used to emit a sound signal, which can be perceived by the microphones and processed by the on-board CPU. In addition to a large number of sensors for perceiving the environment, several sensors provide each s-bot with information about physical contacts, efforts, and reactions at the interconnection joints with other s-bots. These include torque sensors on most joints as well as a traction sensor, which detects the direction and the intensity of the pulling force that the turret exerts on the chassis resulting from the forces applied by other connected s-bots.

7.3 Experiments

By exploiting the swarm-bot robotic platform, we performed a series of experiments, all characterized by a coherent methodological approach. First of all, evolution was always performed in a simulated environment, which was designed to model the relevant features of the s-bot. When required by the experimental setup, the simulation exploited a full 3D physics simulation. This is the case for the experiments presented in section 7.3.1, in which pulling/pushing forces have a fundamental role in the swarm-bot behavior. Otherwise, we employed minimal simulations. In any case, the evolved controllers have been ported to reality to test the viability of the obtained controllers.

All evolutionary experiments share the same methodological approach. The algorithm is run for a fixed number of generations and works on a single population of genotypes. Each genotype encodes the parameters of a single neural network controller. During evolution, a genotype is mapped into a control structure that is cloned and downloaded in all the s-bots taking part in the experiment (i.e., we make use of a homogeneous group of s-bots). Each genotype is evaluated over multiple trials. The fitness of a genotype is the average performance computed over the trials in which the corresponding neural controller is tested. The homogeneous group resulting from a single genotype allows us to simplify the fitness assignment problem. In fact, a single controller is evaluated and selected for the group performance. This group selection also facilitates the evolution of cooperative strategies, given that there is no competition between different individuals in the group.

In the following sections 7.3.1–7.3.4, we present four different experiments performed within the SWARM-BOTS project exploiting the ER approach: coordinated motion and hole avoidance, synchronization, categorization, and self-assembly. In all four sections, we first introduce the scenario in which these experiments have been performed, we discuss the experimental setup, and finally we draw some conclusions about the lesson learned from the study.

7.3.1   Coordinated Motion and Hole Avoidance

The Scenario

For a swarm-bot to move coherently, s-bots need to negotiate a common direction of motion and maintain the group coordination against external disturbances. The coordinated motion of the assembled structure must take into account the variable number of assembled units, as well as a varying topology. Moreover, the swarm-bot's navigation must be efficient with respect to any obstacle and other hazards such as holes and rough terrain, which may be perceived only by a limited subset of the connected s-bots.

Coordinated motion has been widely studied in the literature (Balch and Arkin 1998; Fredslund and Matarić 2002; Quinn et al. 2003; Spector et al. 2005). However, in the swarm-bot case, it takes a different flavor, due to the physical connections among the s-bots, which open the way to study novel interaction modalities that can be exploited for coordination. The experimental scenario can be summarized as follows: at the beginning of a trial, the s-bots start with their chassis oriented in a random direction. Their goal is to choose a common direction of motion on the basis of only the information provided by their traction sensor, and then to move as far as possible from the starting position (Baldassarre et al. 2007). In a different set of experiments, the experimental arena presents holes and open borders, in which a swarm-bot risks remaining trapped. In this case, s-bots must coordinate with the rest of the group to avoid falling (Trianni and Dorigo 2006). Notice that this task is more difficult than it might appear at first sight. First, the group is not driven by a centralized controller (i.e., the control is distributed). Moreover, s-bots cannot use any type of landmark in the environment, such as light sources, or exploit predefined hierarchies between them to coordinate (i.e., there is no "leader robot" that decides and communicates to the other robots the direction of motion of the whole group). Finally, the s-bots do not have a predefined trajectory to follow, nor they are aware of their relative positions or about the structure of the swarm-bot in which they are assembled. As a consequence, the common direction of motion of the group should result from a self-organizing process based on local interactions, which are shaped as traction forces. The problem of designing a controller capable of producing such a self-organized coordination is tackled using feed-forward neural networks synthesized by artificial evolution.

Results Obtained

As mentioned earlier, in order to move in a coordinated way s-bots can rely only on the traction sensor information, which provides a coarse indication of the average direction of motion of the group. By physically integrating the pulling/pushing forces that the connected s-bots produce, the traction sensor provides compact information that can be exploited for coordination. The problem is therefore designing a controller that would let the group self-organize by interacting through physical forces. The results obtained evolving coordinated motion are extremely interesting (Baldassarre et al. 2007). The evolved neural network encodes simple control rules that allow the robots to consistently achieve a common direction of motion in a very short time, and compensate possible misalignments during motion. In general terms, the evolved strategy is based on two feedback loops. Positive feedback makes robots match the average direction of motion of the group, as it is perceived through the traction sensor. Negative feedback makes robots persist in their own direction of motion, but when the traction and motion directions are opposite. Thus the positive feedback allows for a fast convergence toward a common direction of motion, which is stabilized by the negative feedback loop that avoids deadlock conditions.

All this is synthesized in a simple neural network evolved in simulation and tested on real robots (see figure 7.2). The performance of the evolved controllers in terms of robustness, adaptation to varying environmental conditions, and scalability to different number of robots and different topologies is striking, demonstrating how evolution

Figure 7.2
(a) Four real s-bots forming a linear swarm-bot during coordinated motion. (b) A physical swarm-bot while performing hole avoidance. Notice how physical connections among the s-bots can serve as support when a robot is suspended out of the arena, still allowing the whole system to work.

synthesized a very efficient self-organizing behavior for coordinated motion (Baldassarre et al. 2007).

Exploiting a similar setup, we also studied how a swarm-bot can navigate in an arena presenting holes or open borders in which the robots risk remaining trapped (Trianni and Dorigo 2006). In this case, we investigated how the swarm-bot can maintain coordination despite the presence of hazardous situations that are perceived only by a subset of the robots involved. To this purpose, some form of communication may be necessary to the group for a quick reaction. We tested three different communication modalities: (1) direct interactions (DI) through pulling/pushing forces, (2) direct communication (DC), handcrafted as a single-tone signal emitted as a reflex to the perception of the hazard, and (3) direct communication in which signaling was controlled by the evolved neural network (evolved communication, EC). In all cases, the s-bots' motion was controlled by a simple perceptron network similar to the one used for coordinated motion. Additionally, s-bots could use their sensors for perceiving the presence of holes in the ground. In the DC and EC setups, s-bots could also communicate with each other through sound signaling (Trianni and Dorigo 2006).

The results obtained show that it is possible to evolve efficient navigation strategies with each communication paradigm we devised. In the DI setup, when only direct interactions are present, the pulling/pushing forces are sufficient to trigger collective hole avoidance. However, in some cases the swarm-bot is not able to avoid falling because the signal encoded in the traction force produced by the s-bots that perceive the hazard may not be strong enough to trigger the reaction of the whole group. A different situation can be observed in the DC and EC setup, in which direct communication allows a faster reaction of the whole group, as the emitted signal immediately reaches all the s-bots. Therefore, the use of direct communication among the s-bots is particularly beneficial in the case of hole avoidance. It is worth noting that direct communication acts here as a reinforcement of the direct interactions among the s-bots. In fact, s-bots react faster to the detection of the hole when they receive a sound signal, without waiting to perceive a traction strong enough to trigger the hole avoidance behavior. However, traction is still necessary for avoiding the hole and coordinating the motion of the swarm-bot as a whole.

We performed a statistical analysis to compare the three different setups we studied, and the results obtained showed that the completely evolved setup outperforms the setup in which direct communication is handcrafted. This result is in our eyes particularly significant, because it shows how artificial evolution can synthesize solutions that would be very hard to design with conventional approaches. In fact, the most effective solutions discovered by evolution exploit some interesting mechanisms for the inhibition of communication that would have been difficult to devise without any a priori knowledge of the system's dynamics (Trianni and Dorigo 2006).

The Lesson Learned

The experiments performed with coordinated motion and hole avoidance revealed how direct interactions through pulling/pushing forces can be exploited to obtain robust coordination strategies in a swarm-bot. The connections among s-bots in fact represent an important means of transferring information through physical forces. However, exploiting such information is not an easy endeavor if a precise model of the traction sensor is not available. In particular, with respect to the synthesis of self-organizing behaviors, the top-down approach runs into troubles due to the complex dynamical interactions among the system components that can hardly be predicted or modeled. The evolutionary approach, instead, does not need any precise model of the system. It is sufficient to test potential solutions and to compare their performance on the basis of a user-defined metric. With respect to handcrafted solutions, the evolutionary approach can achieve a better performance as it can better exploit all system features, without being constrained by a priori assumptions. This is clear in the hole avoidance experiments, which show how the handcrafted reflex signaling, which seemed perfectly reasonable at first sight, is outperformed by the evolved signaling strategy, which could exploit self-inhibitory mechanisms that are counterintuitive for a "naive" designer.

7.3.2 Synchronization

The Scenario

An important feature of a swarm robotics system is the coordination of the activities through time. Normally, robots can be involved in different tasks, and higher efficiency may be achieved through the synchronization of the activities within the swarm. Synchrony is a pervasive phenomenon: examples of synchronous behaviors can be found in the inanimate world as well as among living organisms (Strogatz 2003). The synchronization behaviors observed in nature can be a powerful source of inspiration for the design of swarm robotic systems, where emphasis is given to the emergence of coherent group behaviors from simple individual rules. Much work takes inspiration from the self-organized behavior of fireflies or similar chorusing behaviors (Holland and Melhuish 1997; Wischmann et al. 2006; Christensen, O'Grady, and Dorigo 2009). Here, we present a study of self-organizing synchronization in a group of robots based on minimal behavioral and communication strategies (Trianni and Nolfi 2009). We follow the basic idea that if an individual displays a periodic behavior, it can synchronize with other (nearly) identical individuals by temporarily modifying its behavior in order to reduce the phase difference with the rest of the group. In this work, the period and the phase of the individual behavior are defined by the sensorimotor coordination of the robot, that is, by the dynamical interactions with the environment that result from the robot embodiment. The studied task requires that each robot in the group display a simple periodic behavior, which should be entrained with the periodic behavior of the other robots present in the arena. The individual periodic behavior consists in oscillations along the y-direction of a rectangular arena (see figure 7.3). Oscillations are possible through the exploitation of a symmetric gradient in shades of gray painted on the ground.

Synchronization of robots movements can be achieved by exploiting a binary, global communication: each robot can produce a continuous tone with fixed frequency and intensity. When a tone is emitted, it is perceived by every robot in the arena, including the signaling one. The tone is perceived in a binary way, that is, either there is someone signaling in the arena, or there is no one. This is a very minimal

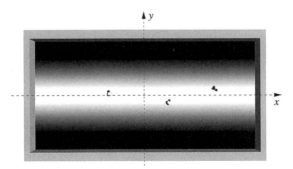

Figure 7.3
Snapshot of a simulation showing three robots in the experimental arena. The dashed lines indicate the reference frame used in the experiments.

communication system for a swarm of robots, which carries no information about the number of signalers, or about their position in the environment. No assumption is made on the way the robots should move on the arena, and on the way they should communicate. All the behavioral rules are designed by the evolution of feedforward neural controllers.

Results Obtained

We performed twenty evolutionary replications, each resulting in the evolution of efficient synchronization behaviors. The individual ability to perform oscillatory movements is based on the perception of the gradient painted on the arena floor, which gives information about the direction parallel to the y-axis and about the point where to perform a U-turn and move back toward the x-axis. The main role of the evolved communication strategy is to provide a coupling between the oscillating s-bots, in order to achieve synchronization: we observed that s-bots change their behavior in response to a perceived communication signal coming from other robots. Recall that the communication signal, being binary and global, does not carry information about either the sender or about its oscillation phase. The reaction to a perceived signal is therefore adapted by evolution to allow the robots to reduce the phase difference between their oscillations, eventually achieving synchronous movements. In summary, the evolved synchronization behaviors are the results of the dynamical relationship between the robot and the environment, modulated through the communicative interactions among robots. No further complexity is required at the level of the neural controller: simple and reactive behavioral and communication strategies are sufficient to implement effective synchronization mechanisms. To better understand the dynamical relationship between individual sensorimotor coordination and communication, we introduced a dynamical system model of the robots interacting with the environment and among each other (Trianni and Nolfi 2009).

This model offers us the possibility to deeply understand the evolved behaviors, both at the individual and collective level, by uncovering the mechanisms that artificial evolution synthesized to maximize the user-defined utility function. We assumed an idealized, noise-free and collision-free environment, and we modeled the s-bot individual behavior as it is produced by the evolved neural network. By coupling the individual behaviors through the communication channel, we could study the effects of perturbations through sound signals over the robot oscillations. We analyzed the different evolutionary runs performed, and we discovered two alternative mechanisms for synchronization. With the modulation mechanism, s-bots synchronize by tuning their oscillatory frequency in response to the perceived communication signal coming from other robots, in order to match the other robots' oscillations. They do so basically by anticipating or delaying the U-turn. With the reset mechanism, s-bots "reset" their oscillation phase by moving to a particular position over the painted gradient, waiting for the other robots to reach a similar position. Qualitatively, similar mechanisms are also observed in biological oscillators. For instance, different species of fireflies present different synchronization mechanisms, based on delayed or advanced phase responses.

Besides studying the synchronization mechanisms, we performed a scalability analysis to test all evolved behaviors with varying group sizes. While scalability is ensured for small groups, we found that physical interactions may prevent the system from scaling to very large number of robots due to the higher probability of performing collision-avoidance maneuvers. Still, the evolved synchronization mechanism scales well if there are no physical interactions. We found that many controllers present perfect scalability, with only a slight decrease in performance due to the longer time required by larger groups to perfectly synchronize. Some controllers, however, present a communicative interference that prevents large groups from synchronizing: the signals emitted by different s-bots overlap in time and are perceived as a fixed signaling pattern. If the perceived signal does not vary in time, it does not bring information to be exploited for synchronization. This problem is mainly due to the global and binary communication form, in which the signal emitted by an s-bot is perceived by any other s-bot anywhere in the arena. Moreover, from the perception point of view, there is no difference between a single s-bot and a thousand signaling at the same time. In order to understand the conditions under which this communicative interference takes place, we again exploited the mathematical model. We found that scalability can be predicted just by looking at the features of the individual behavior: the synchronization behavior scales to any number of robots provided that an s-bot that perceives a communication signal never emits a signal itself. This is a very interesting result, as it directly relates the collective behavior to the individual one, and indicates which are the building blocks for obtaining scalability in the system under study (Trianni and Nolfi 2009).

The Lesson Learned

The synchronization experiments show how temporal coordination can be achieved exploiting simple self-organizing rules. To this purpose, it is not necessary to provide robots with complex behaviors and time-dependent structures. Instead, we show that a minimal complexity of the behavioral and communicative repertoire is sufficient to observe the onset of synchronization. Robots can be described as embodied oscillators, their behavior being characterized by a period and a phase. In this perspective, the movements of an s-bot correspond to advancements of its oscillation phase. Robots can modulate their oscillations simply by moving in the environment and by modifying their dynamical relationship with it. Such modulations are brought forth in response to the perceived communication signals, which also depend on the dynamical relationship between the s-bot and the environment.

In this perspective, the dynamical system analysis proved very useful: we introduced a dynamical system model of the robots interacting with the environment and each other. This model offered us the possibility to deeply understand the evolved behaviors, both at the individual and collective level, by uncovering the mechanisms that artificial evolution synthesized to maximize the user-defined utility function. Moreover, the developed model can be used to predict the ability of the evolved behavior to efficiently scale with the group size. We believe that such predictions are of fundamental importance to quickly select or discard obtained solutions without performing a time-demanding scalability analysis, as well as to engineer swarm robotic systems that present the desired properties. For instance, the knowledge acquired through the performed analysis could be exploited to improve the experimental setup. We have found that the communicative interferences that prevent the group from synchronizing are caused by a communication channel that is neither additive nor local. The locality of communication certainly is an important issue to take into account when studying a realistic experimental setup. Additivity, that is, the capability of perceiving the influence of multiple signals at the same time, is also crucial for self-organizing behaviors. We tested the latter issue, and we discovered that it is sufficient to provide the robots with the average signaling activity of the group to systematically evolve scalable behaviors (Trianni and Nolfi 2011).

7.3.3   Categorization, Integration over Time, and Collective Decisions
The Scenario
A general problem common to biology and robotics concerns the understanding of the mechanisms necessary to decide whether to pursue a particular activity or to give up and perform alternative behaviors. This problem is common to many activities that natural or artificial agents are required to carry out. Autonomous agents may be asked to change their behavior in response to the information gained through repeated interactions with their environment. For example, after various unsuccessful attempts to retrieve a heavy prey, an ant may decide to give up and change its behavior by either cutting the prey or recruiting some nest-mates for collective transport (Detrain and Deneubourg 1997). This example suggests that autonomous agents require adaptive mechanisms to decide whether it is better to pursue solitary actions or to initiate cooperative strategies.

We confronted with the decision-making problem by designing the experimental scenario depicted in figure 7.4. Robots are positioned within a boundless arena containing a light source. Their goal is to reach a target area around the light sources. The color of the arena floor is white except for a circular band around the lamp, within which the floor is in shades of gray. The robots can freely move within the band, but they are not allowed to cross the black edge. The latter can be imagined as an obstacle or a trough that prevents the robot from further approaching the light. The goal of

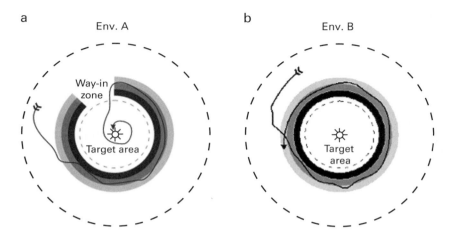

Figure 7.4
Depiction of the task. (a) Env. A is characterized by the way in zone. The target area, centered on the light, is indicated by the dashed circle. (b) In Env. B there is no way in zone and the target area cannot be reached. The continuous lines are an example of a good navigation strategy for one robot.

the experiments is to show that the robots can learn to discriminate between two types of environments. In the first type—referred to as Env. A—the band presents a discontinuity (see figure 7.4a). This discontinuity, referred to as the "way in zone," is a sector of the band in which the floor is white. In the second type—referred to as Env. B—the band completely surrounds the light (see figure 7.4b). The way in zone represents the path along which the robots are allowed to safely reach the light in Env. A. Successful robots should prove capable of performing phototaxis and of moving over the circular band in search for the way in zone, without crossing the black edge. When placed in Env. A, the robots should always reach the target area. When placed in Env. B, on the contrary, the robot should initiate an alternative action, such as signaling or moving away in order to search for other light sources.

Initial experimentation was performed using a single robot controlled by an evolved continuous-time recurrent neural network (CTRNN) (Beer 1995). The results revealed that decision making could be performed by exploiting a temporal cue: the Env. B can be "recognized" by the persistence of a particular perceptual state for the amount of time necessary to discover that there is no way in zone. The flow of time, in turns, can be recognized through the integration of the perceptual information available to the robot. This means that the movements of the robot should bring forth the persistence of a certain perceptual condition, and the discrimination can be made only if the latter is maintained long enough.

We repeated the experiments using two robots having the same sensorimotor capabilities (Ampatzis et al. 2008). Additionally, robots are provided with a communication system similar to the one used in the synchronization experiments: they can emit a single frequency tone that is perceived everywhere in the arena in a binary way. The experiments have been performed by varying the initial position of the two robots, and by rewarding them when they perform antiphototaxis when placed in Env. B. However, no explicit reward was given for communication among the robots. In this way, we aimed at observing whether cooperative communicative behavior could emerge or not.

Results Obtained

Twenty evolutionary simulation runs, each using a different random initialization, were performed for 12,000 generations. Thirteen evolutionary runs produced successful groups of robots: both robots approach the band and subsequently (1) reach the target area through the way in zone in Env. A; (2) leave the band performing antiphototaxis in Env. B. The discrimination between the two environments is possible by exploiting the integration over time and the ability of the leaky integrators that form the robot's neural controller. While moving over the circular band, the s-bot accumulates evidence about the absence of the way in zone. If the latter is found, the integration over time is stopped and the robot continues performing phototaxis. If, instead, the way in zone is not present, after approximately one loop, the robot leaves the band. This evolved behavior closely resembles the one obtained with a single robot. However, a closer look reveals that among the thirteen successful groups, nine make use of sound signaling. In particular, signaling strongly characterizes the behavioral strategies of the groups when they are located in Env. B. In Env. A signaling is, for all these groups, negligible.

Note that the emission of sound is not demanded in order to navigate toward the target and discriminate Env. A from Env. B. Indeed, the task and the fitness function do not require the robots to display signaling behavior. Mechanisms for phototaxis, antiphototaxis, and memory are sufficient for a robot to accomplish the task. In order to reveal the adaptive significance of sound signaling, further tests have been performed.

We looked at the behavior of the robots that emit sound during a successful trial in each type of environment. We recorded the behavior of the robots in both a normal condition and a condition in which the robots cannot hear each other's sounds.

In the normal condition we notice that as soon as one of the robots starts signaling, both robots initiate an antiphototactic movement. But when communication signals are blocked, we notice that each robot initiates antiphototaxis only at the time when it starts emitting its own sound. Sound signaling has therefore the function of stimulating antiphototaxis also for those robots that have not yet gathered enough evidence about the absence of the way in zone.

These results show that most successful strategies employ signaling behavior and communication among the members of the groups. However, communication was not explicitly rewarded: communicating and noncommunicating groups could in principle obtain equal fitness. This means that communication may have other functions that influence its adaptive significance. By looking at the behavior of all successful groups, we discovered that whenever signaling is functionally relevant, robots employ it in Env. B as a self-produced perceptual cue. This cue induces the emitter as well as the other robot of the group to change its behavior from light seeking to light avoiding.

This evidence constrains our investigation on the adaptive significance of sound signaling to two functions: on the one hand, sound is the means by which a robot emitter switches from phototaxis to antiphototaxis. We refer to this as the "solitary" function. On the other hand, sound is the means by which the robot emitter influences the behavior of the other robot. We refer to this as the "social" function. From the data we gathered, it appears that signaling is beneficial mainly because of its "social" function.

The selective advantage of signaling groups is given by the beneficial effects of communication with respect to a robust disambiguation of Env. A from Env. B. The task in fact requires one to find an optimal trade-off between speed and accuracy of the decision.

The beneficial effect of communication corresponds to robust individual decision making and faster group reaction, since signaler and hearer react at the same time. In fact, a robust individual decision requires longer time spent over the circular band to accumulate evidence of the absence of the way in zone, due to the environmental noise that influences the sensors and to the uncertainty of the action outcomes. In total, in those groups in which antiphototaxis is triggered by the perception of sound, a robot that by itself is not ready to make a decision concerning the nature of the environment can rely on the decision taken by the other robot of the group. In average, communication allows the group to accomplish the task earlier, and more reliably. In this way, signaling groups are better adapted to the "danger" of discrimination mistakes in Env. A than are nonsignaling groups, and thus "early" signaling seems to be an issue that has been taken care of by evolution. In fact, once signaling groups evolve, their signaling behavior is refined by categorizing the world later than in the case of nonsignaling groups. This happens in order to ensure that the chances of a potential disadvantage resulting from social behavior are minimized. In other words, the use of communication in a system can also affect aspects of the behavior not directly related to communication (i.e., the process of integration of inputs over time).

The Lesson Learned
The experiments presented in this section show how individual decision making and group behavior can be coevolved to obtain a robust and efficient system. The need to

perform a decision on the basis of information accumulated over time creates a natural trade-off between speed and accuracy. Each s-bot has to resolve a dilemma: to continue searching for the way in zone, or to leave for good? The solution, under normal evolutionary pressures, would be to tune the individual behavior to limit the time spent searching to the minimum. However, the introduction of other robots contemporaneously solving the same task, and the possibility of communication, changes the evolutionary dynamics. By exploiting the information gathered by other robots, it is possible to improve the accuracy of the group decision without reducing the decision speed. This is a relevant fact, which justifies the usage of a collective robotics setup even for those conditions in which it is not explicitly required. Additionally, the exploitation of communicative strategies allows each robot to spread acquired information to the group, and to share information retrieval duties among group members: in fact, as soon as communication is in place, the individual behavior can be refined to exploit the redundancy of the system to the maximum.

7.3.4 Self-assembly and Autonomous Role Allocation

The Scenario

Self-assembly is a ubiquitous process in nature. According to Whitesides and Grzybowski (2002), it is defined as "the autonomous organisation of components into patterns or structures without human intervention." At the nano- or microscopic scale, the interaction among components is essentially stochastic and depends on their shape, structure, or chemical nature. Nature also provides many examples of self-assembly at the macroscopic scale, the most striking being animals forming collective structures by connecting to one another. Individuals of various ant, bee, and wasp species self-assemble and manage to build complex structures such as bivouacs and ladders (Anderson, Theraulaz, and Deneubourg 2002; Hölldobler and Wilson 1978).

As mentioned in section 7.1, the robotics community has been largely inspired from cooperative behavior in animal societies when designing controllers for groups of robots that have to accomplish a given task. In particular, self-assembly provides a novel form of cooperation in groups of robots. However, it is important to notice that some characteristics of the hardware may impose important constraints on the control of the modules of a self-assembling system. As argued by Tuci et al. (2006a), some hardware platforms consist of morphologically heterogeneous modules that can only play a predefined role in the assembly process. In others, the hardware design does not allow, for example, the assembly of more than two modules, or requires extremely precise alignment during the connection phase—that is, it requires a great accuracy. The swarm-bot platform, thanks to its sensors and actuators and its connection apparatus, does not severely constrain the design of control mechanisms for self-assembly. The lack of hardware constraints and the homogeneity of the robots require that self-assembly be achieved through a differentiation of roles, resulting in the definition of an s-bot gripper (i.e., the robot that makes the action of gripping) and an s-bot grippee (i.e., the robot that is gripped). In work carried out within the SWARM-BOTS project by using control design techniques other than ER, the s-bot gripper/s-bot grippee differentiation was either predefined (Groß et al. 2006) or based on stochastic events and a complex communication protocol (O'Grady et al. 2005). Thanks to the use of ER we designed control strategies for real assembling robots that are not constrained by either morphological or behavioral heterogeneities introduced by the hardware and control method, respectively (see Ampatzis et al. 2009, for details). Instead of a priori defining the mechanisms leading to role allocation and self-assembly, ER allowed us to let behavioral heterogeneity emerge from the interaction among the system's homogeneous components. Moreover, coordination and cooperation in self-assembly between physical robots is achieved without requiring explicit signaling of internal states, as assumed, for example, in Groß et al. 2006.

Self-assembly is studied in a scenario in which two s-bots are positioned in a boundless arena at a distance randomly generated in the interval [25 cm,30 cm], and with predefined initial orientations. The robots are required to approach each other and to physically assemble through the gripper. The agents perceive each other through their omni-directional camera mounted on the turret, which returns rough information about robot distance and orientation. We also make use of the optical barrier mounted on the gripper, which informs a robot about the presence of an object between the gripper claws. The agent controller is composed of a CTRNN, whose control parameters are evolved through a rank-based evolutionary algorithm.

Results Obtained

The results of this work prove that dynamical neural networks shaped by evolutionary computation techniques directly controlling the robots' actuators can provide physical robots with all the required mechanisms to autonomously perform self-assembly. Owing to the ER approach, the assembly is initiated and regulated by perceptual cues that are brought forth by the homogeneous robots through their dynamical interactions. Moreover, in spite of the system being homogeneous, role allocation—in other words, who is the s-bot gripper and who is the s-bot grippee—is successfully accomplished by the robots through an autonomous negotiation phase between the two s-bots, as confirmed by our behavioral analyses (see figure 7.5). We observed that role allocation unfolds in time during the entire duration of a trial.

Whenever the two robots have different initial perceptions, the role that each s-bot assumes can be predicted knowing the combination of the initial relative orientations of the robots. In other words, the combination of relative orientations leads to a pattern of interactions among the robots with a predictable outcome, from the observer point of view. However, a robot has no such information. Perceiving the other robot at a specific distance and orientation does not inform a robot about the role it will

Figure 7.5
Snapshots from a successful trial: (a) initial configuration, (b) starting phase, (c) role allocation phase, (d) gripping phase, (e) success (grip).

assume at the end of the trial. In summary, whenever the initial orientations are asymmetrical, robots engage in a role negotiation phase, and the dynamical system composed of the two interacting robots almost always converges at the same final condition, which depends only on the initial conditions.

In those cases in which the robots start with an identical perception, symmetry does not hinder the robots from autonomously allocating different roles to successfully accomplish their goal. The robots engage in a dynamical interaction, which eventually leads to a role assignment. However, in this case it is not possible to predict the outcome of the role allocation process: both robots have 50 percent probability of assuming the s-bot gripper or the s-bot grippee role. Post-evaluation tests have shown that the random noise inherent in the system is the causal factor that drives the system through sequences of actions that turn out to be successful. In other words, the dynamical system composed by the two interacting robots starts from an unstable equilibrium point, from which it can converge at either stable condition, that is, at one of the two alternative role allocations. It is important to notice that the symmetry breaking is performed by exploiting randomness present in the system, which is amplified by the neural controllers as a result of the evolutionary optimization.

Finally, tests with real robots revealed that the evolved mechanisms proved to be robust with respect to changes in the color of the light displayed by the LEDs. Furthermore, the self-assembling robotic system designed by using ER techniques exhibits recovery capabilities that could not be observed during the artificial evolution and that were not coded or foreseen by the experimenter (Ampatzis et al. 2009). Such a feature in our case comes for free, while in the case of Groß et al.'s experiments (2006) a recovery mechanism had to be designed as a specific behavioral module to be activated every time the robots failed to achieve assembly.

The Lesson Learned

The main contribution of this work lies in the design of control strategies for real assembling robots that are not constrained by morphological or behavioral heterogeneities introduced by the hardware and control method, respectively. Contrary to the modular or hand-coded controllers described by Groß et al. (2006) and O'Grady et al. (2005), the evolutionary robotics approach did not require the experimenter to make any a priori assumption concerning the roles of the robots during self-assembly (i.e., either s-bot gripper or s-bot grippee) or about their status (e.g., either capable of moving or required not to move). We showed with physical robots that coordination and cooperation in self-assembly do not require explicit signaling of internal states, as assumed, for example, by Groß et al. (2006). In other words, we present a setup that requires minimal cognitive and communicative capacities on behalf of the robots. The absence of a priori assumptions allows evolution to exploit the dynamical interaction among the robots to produce an autonomous role allocation mechanism. This can be considered an example of a self-organizing system close to a bifurcation point, in which the random fluctuations of the system are amplified to let the system overcome the impasse given by symmetric starting conditions and converge toward a desired solution.

7.4   Discussion

The experiments presented in section 7.3 are representative of a coherent theoretical and methodological approach to the synthesis of self-organizing behaviors for a swarm robotics system. What are the limits of this approach? The main problem to deal with is the evolvability of the system related to the scaling in complexity of the collective behavior. By practicing with evolutionary swarm robotics, it appears rather easy to evolve self-organizing behaviors in which the system achieves and maintains a certain spatiotemporal pattern. For instance, coordinated motion of the swarm-bot and synchronization are not particularly difficult to evolve (e.g., they require few generations, and successful controllers are almost always obtained), once a suitable experimental setup has been defined (see sections 7.3.1 and 7.3.2).

On the one hand, this is justified by the simplicity of the neural controller and the rather limited number of free parameters that need to be optimized by the evolutionary machinery. On the other hand, the quality of the interactions among the robots contains in itself part of the solution to the self-organization problem.

In the whole, simple controllers and well-defined interactions represent a perfect starting point for the evolution of self-organizing behavior. As a matter of fact, in similar conditions successful behaviors are systematically obtained in all evolutionary runs.

However, the situation is slightly different when evolution must produce self-organizing systems close to a bifurcation point, in which multiple solutions are possible as a result of the interactions, feedback loops, and randomness of the system. This is the case of the categorization experiment, in which robots had to take a collective decision (section 7.3.3), and of the self-assembly experiment, in which complementary roles needed to emerge from the robot-robot interactions and the amplification of random fluctuations of the system (section 7.3.4). In similar conditions, evolvability is limited by the need to contemporaneously evolve different behavioral traits, and by the presence of multiple stable conditions, which create local optima in which evolution may remain trapped. In the experiments we performed, many generations were necessary to find a suitable solution. Also, the success rate was never close to 100 percent, and some evolutionary runs resulted in partial solutions of the problem. The evolution of communication raises a similar problem, requiring evolution of both the signal and the response to the signal, which individually may be counteradaptive or neutral with respect to the devised fitness function (see section 7.3.3).

The experiments presented in section 7.3.3 are interesting also from a different point of view, that is, the influence that the individual behavior has on the evolution of the group behavior. Here, we can distinguish between two organizational levels: (1) the individual level, in which sensorimotor coordination and integration over time support the decision making, and (2) the collective level, in which information spreading through communication leads to increased group efficiency. We believe that future directions in evolutionary swarm robotics should focus on systems characterized by multiple levels of organization. More complex self-organizing behaviors can be obtained through a layered evolution that proceeds through individual sensorimotor coordination, individual categorization abilities, and communication and exploitation of the social environment, aiming at some collective intelligence. As experienced in our experiments, each different level of organization is supported by the lower levels, and in turns influences their dynamics. In a swarm robotics scenario, the influences of the higher organizational level on the lower ones could be exploited to simplify the individual behavior in favor of more robust, collective solutions. Brought to the limit, each robot in the swarm could behave as a neuron-like device that can move in the environment and interact, physically or through communication, with neighboring robots, while the swarm brings forth complex processes as a whole. In this respect, we believe that the cognitive abilities of swarms should be studied and compared with those observed in the vertebrate brain, in the attempt to find the common mechanisms that underlie cognition. In this respect, robotics models of swarm behavior may represent extremely powerful tools for the study of swarm cognition (Trianni et al. 2011).

Another possible direction in the study of evolutionary swarm robotics concerns the exploitation of heterogeneous swarms, in which different types of robots are organized in swarms, which cooperate for a collective goal. We investigated swarms of heterogeneous robots within the project Swarmanoid,[1] in which three types of robots have been studied: eye-bots, foot-bots, and hand-bots. Eye-bots are robots specialized in sensing and analyzing the environment from a high position to provide an overview that foot-bots or hand-bots cannot have. Eye-bots fly or are attached to the ceiling. Hand-bots are specialized in moving and acting in a space zone between the one covered by the foot-bots (the ground) and the one covered by the eye-bots (the ceiling). Hand-bots can climb vertical surfaces. Foot-bots are specialized in moving on rough terrain and transporting either objects or other robots. They are based on the s-bot platform, and extend it with novel functionalities. The combination of these three types of autonomous agents forms a heterogeneous swarm robotic system that is capable of operating in a 3D space.

Generally speaking, dealing with heterogeneity in a collective robotics setup often leads to specialization and teamwork: the task is broken down on the basis of the different robots available, and roles are assigned correspondingly. With heterogeneous swarms, the redundancy of the system opens the way to various scenarios. On one extreme, the classical scenario accounts for different swarms that specialize in particular subtasks, and are loosely coupled. For instance, a swarm of eye-bots is responsible of locating areas of particular interest, such as areas that contain objects to be retrieved. The eye-bots direct the action of a swarm of foot-bots, which collectively retrieve such objects. On the other extreme, robots can form a swarm of homogeneous entities, where each entity is a small, heterogeneous, tightly cooperating team. For instance, two or three foot-bots can self-assemble to transport a single hand-bot, thereby creating a small team, which can coordinate its activities within a swarm of similar foot-bot/hand-bot teams. Between these two extreme scenarios, there can be an infinite blend of possibilities for cooperating heterogeneous swarms. In this respect, ER can give a strong contribution to define the individual behaviors, and shape the self-organization of the heterogeneous swarm. In particular, ER can be exploited to define the behavior of the heterogeneous robots by evolving one controller for each robot type. An alternative, interesting scenario consists of synthesizing homogeneous controllers for heterogeneous robots, in which the controller adapts to the dynamics of the robot on which it is downloaded without a priori knowledge of its type. We performed preliminary studies by evolving controllers for a heterogeneous group of three simulated robots (Tuci et al. 2008). The agents are required to cooperate in order to avoid collisions when approaching a light source. The robots are morphologically different: two of them are equipped with infrared sensors, one with light sensors. Thus, the two morphologically identical robots should take care of obstacle avoidance, while the other one should take care of phototaxis. Since all the agents can emit and perceive sound, the group's coordination of actions is based on acoustic communication. The results of this study are a "proof-of-concept": they show that dynamic artificial neural networks can be successfully synthesized by artificial evolution to design the neural mechanisms required to under pin the behavioral strategies and adaptive communication capabilities demanded by this task. Thus, ER represents a promising method that should be considered in future research works dealing with the design of homogeneous controllers for groups of heterogeneous cooperating and communicating robots.

In conclusion, based on the results obtained in past research and on the prospect of future achievements, we believe that the bidirectional influence arrow connecting ER and swarm robotics can be enforced in both directions. ER can offer swarm robotics a bias-free method to automatically obtain robust and sophisticated control structures that exploit aspects of the experimental setup not always evident a priori to the experimenter. Equally, swarm robotics can broaden the horizons of ER beyond the current limits. In our opinion, the swarm cognition approach and studies with heterogeneous swarms are two of the most promising directions.

Note

1. A project funded by the Future and Emerging Technologies program of the European Community, under grant IST-022888.

References

Ampatzis, C., E. Tuci, V. Trianni, A. L. Christensen, and M. Dorigo. 2009. Evolving autonomous self-assembly in homogeneous robots. *Artificial Life* 15 (4): 465–484.

Ampatzis, C., E. Tuci, V. Trianni, and M. Dorigo. 2008. Evolution of signaling in a multi-robot system: Categorization and communication. *Adaptive Behavior* 16 (1): 5–26.

Anderson, C., G. Theraulaz, and J.-L. Deneubourg. 2002. Self-assemblages in insect societies. *Insectes Sociaux* 49 (2): 99–110.

Balch, T., and R. C. Arkin. 1998. Behavior-based formation control for multirobot teams. *IEEE Transactions on Robotics and Automation* 14 (6):926–939.

Baldassarre, G., V. Trianni, M. Bonani, F. Mondada, M. Dorigo, and S. Nolfi. 2007. Self-organised coordinated motion in groups of physically connected robots. *IEEE Transactions on Systems, Man, and Cybernetics. Part B, Cybernetics* 37 (1): 224–239.

Beckers, R., O. E. Holland, and J.-L. Deneubourg. 1994. From local actions to global tasks: Stigmergy and collective robotics. In *Proceedings of the 4th International Workshop on the Synthesis and Simulation of Living Systems (Artificial Life IV)*, ed. R. A. Brooks and P. Maes, 181–189. Cambridge, MA: MIT Press.

Beer, R. D. 1995. A dynamical systems perspective on agent-environment interaction. *Artificial Intelligence* 72:173–215.

Bonabeau, E., M. Dorigo, and G. Theraulaz. 1999. *Swarm Intelligence: From Natural to Artificial Systems*. New York: Oxford University Press.

Camazine, S., J.-L. Deneubourg, N. Franks, J. Sneyd, G. Theraulaz, and E. Bonabeau. 2001. *Self-Organization in Biological Systems*. Princeton, NJ: Princeton University Press.

Christensen, A. L., R. O'Grady, and M. Dorigo. 2009. From fireflies to fault-tolerant swarms of robots. *IEEE Transactions on Evolutionary Computation*, Special Issue on Swarm Intelligence 13 (4): 754–766.

Couzin, I. D. 2007. Collective minds. *Nature* 455 (7129): 715–715.

Couzin, I. D., and J. Krause. 2003. Self-organization and collective behavior of vertebrates. *Advances in the Study of Behavior* 32:1–75.

Detrain, C., and J.-L. Deneubourg. 1997. Scavenging by pheidole pallidula: A key for understanding decisionmaking systems in ants. *Animal Behaviour* 53:537–547.

Dorigo, M., V. Trianni, E. S. Sahin, R. Groß, T. H. Labella, G. Baldassarre, S. Nolfi, J-L. Deneubourg, F. Mondada, D. Floreano, and L. M. Gambardella. 2004. Evolving self-organizing behaviors for a swarm-bot. *Autonomous Robots* 17 (2–3): 223–245.

Franks, N. R., S. C. Pratt, E. B. Mallon, N. F. Britton, and D. J. T. Sumpter. 2002. Information flow, opinion polling and collective intelligence in house-hunting social insects. *Philosophical Transactions of the Royal Society of London. Series B, Biological Sciences* 357 (1427): 1567–1583.

Fredslund, J., and M. J. Matarić. 2002. A general algorithm for robot formations using local sensing and minimal communication. *IEEE Transactions on Robotics and Automation* 18 (5): 837–846.

Groß, R., M. Bonani, F. Mondada, and M. Dorigo. 2006. Autonomous self-assembly in swarm-bots. *IEEE Transactions on Robotics* 22 (6): 1115–1130.

Holland, O., and C. Melhuish. 1997. An interactive method for controlling group size in multiple mobile robot systems. In *Proceedings of the 8th International Conference on Advanced Robotics (ICAR '97)*, 201–206. Piscataway, NJ: IEEE Press.

Holland, O., and C. Melhuish. 1999. Stigmergy, self-organization, and sorting in collective robotics. *Artificial Life* 5 (2): 173–202.

Hölldobler, B., and E. O. Wilson. 1978. The multiple recruitment systems of the African weaver ant, Œcophylla longinoda (latreille) (hymenoptera: formicidae). *Behavioral Ecology and Sociobiology* 3:19–60.

Jones, C. V., and M. J. Matarić. 2006. Behavior-based coordination in multi-robot systems. In *Autonomous Mobile Robots: Sensing, Control, Decision-Making, and Applications*, ed. S. S. Ge and F. L. Lewis, 549–569. Boca Raton, FL: CRC Press.

Krieger, M. J. B., J.-B. Billeter, and L. Keller. 2000. Ant-like task allocation and recruitment in cooperative robots. *Nature* 406:992–995.

Kube, C. R., and H. Zhang. 1993. Collective robotics: From social insects to robots. *Adaptive Behavior* 2 (2): 189–219.

Martinoli, A., A. J. Ijspeert, and F. Mondada. 1999. Understanding collective aggregation mechanisms: From probabilistic modeling to experiments with real robots. *Robotics and Autonomous Systems* 29:51–63.

Mondada, F., G. C. Pettinaro, A. Guignard, I. V. Kwee, D. Floreano, J.-L. Deneubourg, S. Nolfi, L. M. Gambardella, and M. Dorigo. 2004. SWARM-BOT: A new distributed robotic concept. *Autonomous Robots* 17 (2–3): 193–221.

O'Grady, R., R. Grofl, F. Mondada, M. Bonani, and M. Dorigo. 2005. Self-assembly on demand in a group of physical autonomous mobile robots navigating rough terrain. In *Proceedings of the 8th European Conference on Artificial Life (ECAL'05)*, LNCS 3630, 272–281. Berlin: Springer-Verlag.

Quinn, M., L. Smith, G. Mayley, and P. Husbands. 2003. Evolving controllers for a homogeneous system of physical robots: Structured cooperation with minimal sensors. *Philosophical Transactions of the Royal Society of London, Series A: Mathematical, Physical and Engineering Sciences* 361 (1811): 2321–2343.

Spector, L., J. Klein, C. Perry, and M. Feinstein. 2005. Emergence of collective behavior in evolving populations of flying agents. *Genetic Programming and Evolvable Machines* 6 (1): 111–125.

Strogatz, S. H. 2003. *Sync: The Emerging Science of Spontaneous Order*. New York: Hyperion Press.

Sumpter, D. J. T. 2006. The principles of collective animal behaviour. *Philosophical Transactions of the Royal Society of London: Series B* 361:5–22.

Trianni, V., and M. Dorigo. 2006. Self-organisation and communication in groups of simulated and physical robots. *Biological Cybernetics* 95:213–231.

Trianni, V., and S. Nolfi. 2009. Self-organising sync in a robotic swarm: A dynamical system view. *IEEE Transactions on Evolutionary Computation*, Special Issue on Swarm Intelligence 13 (4): 722–741.

Trianni, V., and S. Nolfi. 2011. Engineering the evolution of self-organising behaviours in swarm robotics: A case study. *Artificial Life* 17 (3): 183–202.

Trianni, V., S. Nolfi, and M. Dorigo. 2008. Evolution, self-organisation and swarm robotics. In *Swarm Intelligence. Introduction and Applications, Natural Computing Series*, ed. C. Blum and D. Merkle, 163–192. Berlin, Germany: Springer-Verlag.

Trianni, V., E. Tuci, K. M. Passino, and J. A. R. Marshall. 2011. Swarm cognition: An interdisciplinary approach to the study of self-organising biological collectives. *Swarm Intelligence* 5 (1): 3–18.

Tuci, E., C. Ampatzis, F. Vicentini, and M. Dorigo. 2008. Evolving homogeneous neurocontrollers for a group of heterogeneous robots: Coordinated motion, cooperation, and acoustic communication. *Artificial Life* 14 (2): 157–178.

Tuci, E., R. Grofl, V. Trianni, M. Bonani, F. Mondada, and M. Dorigo. 2006a. Cooperation through self-assembling in multi-robot systems. *ACM Transactions on Autonomous and Adaptive Systems* 1 (2): 115–150.

Tuci, E., R. Grofl, V. Trianni, F. Mondada, M. Bonani, and M. Dorigo. 2006b. Cooperation through self assembling in multi-robot systems. *ACM Transactions on Autonomous and Adaptive Systems* 1 (2): 115–150.

Whitesides, G., and B. Grzybowski. 2002. Self-assembly at all scales. *Science* 295:2418–2421.

Wischmann, S., M. Huelse, J. F. Knabe, and F. Pasemann. 2006. Synchronization of internal neural rhythms in multi-robotic systems. *Adaptive Behavior* 14 (2): 117–127.

8   Evolution of Communication in Robots

Joachim de Greeff and Stefano Nolfi

8.1   Introduction

During the last ten years, the attempt to study the evolution of communication and language through computational and robotic models has attracted the attention of an increasing number of researchers (for a review, see Cangelosi and Parisi 2002; Kirby 2002; Steels 2003; Wagner et al. 2003; Nolfi 2005; Nolfi and Mirolli 2010). Indeed, the study of how populations of artificial agents that are embodied and situated can autonomously develop communication skills and a communication system while they interact with a physical and social environment presents two important advantages with respect to experimental methods: (1) it allows researchers to study how communication signals are grounded in agents' nonsymbolic sensorimotor experiences, and (2) it allows researchers to come up with precise and operational models of how communication skills can originate and how established communication systems can evolve and adapt to variations of the physical and social environment.

Within this area, evolutionary robotics (ER) can provide a key contribution because some of its foundational features differentiate it from other alternative learning methods: the fact that fine-grained characteristics that regulate how the robots interact with the physical and social environment can be encoded into free parameters; and the fact that variations can be retained or discarded on the basis of their affect at the level of the global behavior exhibited by the robot/robots (Nolfi 2009). These features, in fact, allow the experimenter to reduce the number of characteristics that are predetermined and fixed to the minimum and leave the robots free to determine how to solve the adaptive problem.

These features enable us to study whether and how communication can emerge in populations of individuals that are not rewarded directly for communicating. Moreover, they allow us to study the role of the coadaptation of behavioral and communication skills (Nolfi 2005) which, as we will show in the following sections, represents an essential prerequisite for the emergence and complexification of robots' communication skills.

In this chapter we describe an experimental scenario (section 8.2) that is simple enough to be analyzed systematically, but that at the same time includes all the elements necessary to investigate important questions concerning the evolution of communication: What are the conditions that might lead to the evolution of communication skills in a population of initially noncommunicating robots? What is the relation between agents' communicative and noncommunicative behaviors and between different communication modalities (e.g., implicit and explicit communication)? How does the "meaning" of signals originate and evolve and how is this grounded (Harnad 1990) in agents' sensory experience? The key aspects of the chosen scenarios are (1) the fact that the task/environment allows qualitatively different solutions, (2) the fact that the robots are provided with a sensorimotor system that allows them to interact/communicate through different modalities, and (3) the fact that the evolving robots are not rewarded for communicating and are left free to determine how they react to sensory states and sequences of sensory states.

In section 8.3 we present the results of these experiments. Analysis of these results might allow us to generate new data that can partially compensate for the paucity of empirical data caused by the fact that language and communication do not leave direct traces in fossil records. As we will see, the analyses of these synthetic experiments provide hints for confirming or disconfirming existing theories on the evolution of communication and language as proposed by evolutionary biologists, and for formulating new theoretical explanations.

Finally, in section 8.4 we discuss some of the implications of the evolution of cooperative behaviors in evolutionary robotics experiments, with particular reference to the role that sociality might have in the manifestation of open-ended evolutionary processes.

8.2   Experimental Setup

The experimental setup involves two wheeled robots situated in an arena containing two target areas (figure 8.1) that are evolved for being concurrently located in the two target areas and for switching areas as often as possible. The characteristics of the task/environment have been chosen to identify a situation in which the robots should coordinate/cooperate to solve their adaptive problem. In the following subsections we describe the characteristics of the environment, of the robots' body and neural controller, and of the evolutionary algorithm.

8.2.1   The Environment and the Robots

The environment consists of an arena of either 110×110 or 150×150 cm surrounded by walls and containing two target areas with a diameter of 34 cm placed on two randomly selected but non-overlapping positions inside the arena. The floor of the

Figure 8.1
Left: The environment and the robots. The two circular areas of the environment colored in black and white represent the two target areas. Right: The e-puck robotic platform including the ground sensor board and a strip of red paper around the top part of the body that allows for easier visual recognition.

arena and the walls are gray. The two circular portions of the arena corresponding to the two target areas are colored black and white, respectively.

The robotic platform consists of two e-Puck robots (Mondada and Bonani 2007) equipped with the ground sensor-board extension. The robots, which have a diameter of 7.5 cm, are equipped with two motors that control the two corresponding wheels, eight infrared proximity sensors located around the robot's body, three infrared sensors placed on the frontal side of the robot and oriented toward the ground, a VGA camera with a field of view of 36 degrees pointing in the direction of forward motion, and a wireless Bluetooth interface that can be used to send and receive signals to and from other robots. The body of the robot has been covered with a circular strip of red paper to allow robots to detect the presence of another robot in their field of view.

Signals consist of single floating-point values in the range [0.0,1.0], which are transmitted and received through the Bluetooth connection. Each time step both robots emit a signal and detect the signal produced by the other robot. For more details, see De Greef and Nolfi 2010).

8.2.2 The Neural Controller
The neural controller of each robot is provided with seventeen sensory neurons, four internal neurons with recurrent connections, and three motor neurons. The internal neurons receive connections from the sensory neurons and from themselves. The motor neurons receive connections from both the sensory and the internal neurons (figure 8.2).

The sensory layer consists of eight neurons that encode the state of the eight corresponding infrared sensors, three neurons that encode whether the robot detects

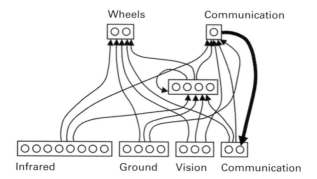

Figure 8.2
The architecture of the robots' neural controller. The lower, middle, and top layers indicate sensory, internal, and motor neurons, respectively. Thin arrows indicate connections. Thick arrows indicate that the state of the communication motor neuron at time t is copied into the state of a sensory neuron at time $t + 1$.

another robot in its field of view and the angular offset of the detected robot on the left or right side of the camera, two neurons that binary encode whether the ground sensor of the robot detects a white or black target area, two neurons that encode the previous state of the ground sensors, and two signal sensors that encode the signal received from the other robot and the signal produced by the robot itself in the previous time step.

The motor layer includes two neurons that encode the desired speed of the two corresponding wheels and one neuron that encodes the value of the signal produced by the robot.

The state of sensory, internal, and motor neurons are updated every 100 ms (i.e., each time step lasts 100 ms). The internal neurons consist of leaky integrator neurons that hold a certain amount of activation from the previous time step and in which the effect of the previous state on their current state is determined by a time-constant parameter. The motor neurons consist of standard sigmoid units. For more details, see De Greeff and Nolfi 2010.

The type and number of sensors and actuators and the way in which the information extracted by the robots' sensors is encoded into the sensory neurons has been chosen to allow the robots to have a potentially rich interaction with their physical and social environment, while keeping the number of free parameters as low as possible.

More specifically, concerning communication, the possibility for the robots to perceive each other potentially allows the development of communication skills, that is, the development of an ability to react to stimuli produced by other individuals in functional ways, or the development of an ability to modify the perceptual environment of the other individuals in functional ways, or both. Moreover, the possibility for robots to influence each other through different modalities (i.e., radio, vision, and infrared) potentially allows the robots to exploit both: (a) *implicit* communication forms, in which the robots develop an ability to react appropriately to the perceptual stimuli that are produced by the other individuals spontaneously, and (b) *explicit* communication forms, in which the robots develop an ability to functionally shape the way in which they affect the perceptual environment of the other individuals. For example, the possibility for the robots to visually detect the presence and the relative position of the other robot might allow the development of implicit communication forms that react to such stimuli in a functional manner with respect to the task/environment. And the possibility for the robots to vary the radio signal produced in different robot/environmental circumstances allows them to develop explicit communication forms in which both the signal produced and detected and the reaction to such signals have been adapted.

For the sake of simplicity, from now on we will use the term "implicit signal" to indicate the signal that is generated by the actual physical position of the robots and that is detected by other robots through their visual and infrared sensors. We will use the term "explicit signal" to indicate the signal produced by a robot and received by other robots through the wireless connection (providing that robots do not always produce the same signal). This is justified by the fact that in this particular experimental setup, the robot can shape the latter stimuli, but not the former, during the adaptive process. We should bear in mind, however, that there are no straightforward ways to formally distinguish between implicit and explicit signals. For example, as we will discuss, the perceptual stimuli generated by the physical position of the robots also can be partially shaped in an adaptive way by the robots themselves through modification of the robots motor behaviors. Finally, the possibility to interact through different communication channels might lead to the development of communication forms that are based on a combination of implicit and explicit signals.

The four sensors that encode both the current and the previous state of the ground sensors allow the robots to easily recognize whether they are or were recently located in one of the target areas.

Finally, leaky internal neurons with recurrent connections allow the evolving robots to integrate sensorimotor information through time (e.g., to detect the duration of a given sensory state) or to remember and eventually communicate previously experienced sensory states, or both (Nolfi and Marocco 2001; Beer 2003). In other words, the characteristics of the neural controllers potentially allow the robots to extract and communicate information that is not currently available through their sensors.

8.2.3 The Evolutionary Algorithm

An evolutionary technique is used to set the free parameters of the robots' neural controller (Nolfi and Floreano 2000). The initial population consists of 100 randomly generated genotypes that encode the connection weights, the biases and time constants of 100 corresponding neural controllers (each parameter is encoded by 8 bits and normalized in the range [–5.0,+5.0] in the case of connection weights and biases and in the range [0.0,1.0] in the case of time constants). Each genotype is translated into two identical neural controllers that are embodied in two corresponding robots situated in the environment (i.e., teams are homogeneous). The twenty best genotypes of each generation are allowed to reproduce by generating five copies each, with 2 percent of their bits replaced with a new randomly selected value. The evolutionary process lasts 1,000 generations (i.e., the process of testing, selecting, and reproducing robots is iterated 1,000 times). The experiment is replicated ten times for each of the two experimental conditions (smaller and larger environment).

Each team of two robots is allowed to "live" for twenty trials, lasting 200 seconds each (i.e., 2000 time steps of 100 ms each). However, if a collision occurs a trial is immediately terminated. At the beginning of each trial the position of the two target areas and the position and orientation of the robots are assigned randomly.

Each team of evolving robots scores 1 point every time the two robots occupy the two different target areas for the first time during a trial or after a switch (i.e., after the robot that previously occupied the white target area moves to the black target area and vice versa). The total performance of a team (fitness) consists of the average number of points scored during the twenty trials.

The robots' neural controllers are evolved in simulation and the best evolved neural controllers have been post-evaluated on hardware (i.e., have been downloaded on the two robots situated in the physical environment).

Before concluding the description of the experimental setup it is important to clarify which characteristics are predetermined by the experimenter and fixed, and which characteristics are unspecified and are left free to vary during the evolutionary process.

One predetermined aspect is that the experimental scenario involves a full cooperative situation. This is due to the fact that the two robots forming a team have the same genetic characteristics and that selection operates on the basis of the performance of the team and not on the performance of a single individual (Floreano et al. 2007). A second predetermined aspect is constituted by the fitness function that is used to select the best individuals. The adaptive task faced by the robots, however, is only partially predetermined since on the one hand it depends on fixed aspects like the fitness function and characteristics of the robots' body and of the environment, but on the other hand also depends on the characteristics of the social environment (i.e., the other robots' behavior), which is not predetermined and varies during the evolutionary process. The development of new behavioral and communication skills modifies the social environment of the robots themselves. These modifications, in turn, modify the adaptive landscape of the robots. Indeed, as we will discuss in section 8.4, modifications of robots' behavior and communication skills may create the adaptive condition for the emergence of new skills.

The motor and communicative behaviors exhibited by the robots are not predetermined since the way in which a robot reacts to any given sensory state or sequence of sensory states depends on the free parameters that are encoded in the genome of the population and are subjected to variations. Indeed, as we will see, evolving robots are left free to determine the characteristics of their behavior within a large space of different behavioral solutions. More specifically, concerning robots' motor behavior, the robots are free to determine the number and the type of elementary behaviors that they display and the way in which these behaviors are combined and arbitrated. Concerning robots' communicative behaviors, evolving robots are left free to determine how to use the information that has a communicative value from the physical and social environment, how many different signals they will produce, in which agent/environmental context each signal will be produced, and what will be the motor and communicative effects of the explicit and implicit signals that are detected. Finally, evolving robots are free to co-adapt their motor and communicative behaviors.

The theoretical approach and the methodology followed in this chapter are in line with the work of Di Paolo (1997, 2000); Quinn (2001); Quinn et al. (2003); Baldassarre, Nolfi, and Parisi (2003); Trianni and Dorigo (2006); Marocco and Nolfi (2007); and Williams, Beer, and Gasser (2008). However, the experimental scenario proposed here is more advanced than in the experimental works mentioned previously with respect to the following aspects (or with respect to the possibility to study the following aspects in combination): (1) the complexity of the task that enable us to study how several behavioral and communication skills are developed and co-adapted during the evolutionary process; (2) the richness of the agents' sensorimotor system that supports, for example, the exploitation of both explicit and implicit communication; and (3) the validation of the results obtained in simulation in hardware.

8.3   Results

The analysis of the results obtained in different replications of the experiment and in different experimental conditions indicates that the robots solve the problem through qualitatively different strategies by exploiting the possibility to communicate through explicit and implicit signals (section 8.3.1). The analysis of the best solutions indicates that evolving robots display rather rich behavioral and communication skills, including: the ability to access/generate information that has a communicative value, the ability to produce different signals encoding useful regularities, and the ability to react appropriately to explicit and implicit signals, by also regulating the reaction on the basis of the context in which signals are detected (section 8.3.2). The analysis of the evolutionary development of the best replications sheds light on how signals/ meanings originate and how robots' behavioral and communication skills progressively complexify as a result of an incremental process. New skills are added on top of previously developed skills, which thus represent a prerequisite for the development and the exhibition of the new skills and which are retained during the successive course of the evolutionary process (see end of section 8.32 and section 8.4). The analysis of evolved behavioral and communication skills and their origin also provides insights on the mechanisms that allow evolving robots to solve the problem resulting from the need to develop an ability to produce useful signals and to react to signals appropriately at the same time (section 8.4).

8.3.1   Performance and Evolved Strategies

By analyzing the results obtained at the end of the evolutionary process for different replications of the experiment and for different experimental conditions (i.e., 110×110 and 150×150 cm arenas) we observed that evolved robots display an ability to be concurrently located in the two areas and to switch area several times in the case of the best replications and only a few times in the case of the worst replications. More precisely, the mean number of switches made (+1 point for finding the two target areas for the first time) is 10.035 and 4.680 for the best individuals evolved and tested in the 110×110 and 150×150 environments, respectively.

Evolved robots exploit the possibility to communicate through explicit signals in most of the replications. Indeed, by analyzing the variance of the performance obtained in a standard condition and a control condition in which the evolving robots are forced to always produce a 0.0 signal, we observed that the ability to vary the explicit signals significantly affects the overall performance of the robots (Kruskal-Wallis test, $p < 0.0005$ for both the experiments performed in the small and large arenas). Evolved robots also rely on implicit communications in most of the cases, as we will illustrate.

The visual inspection of the fittest evolved solutions indicates that they can be grouped in two qualitatively different strategies. In both strategies, the robots initially display an exploration behavior that allows them to enter the two target areas (one robot per area) and then display a series of target-switching behaviors in which each robot exploits the information provided by the other individual to navigate directly toward the other target area. The first strategy (that will be called "symmetrical strategy" from now on and that corresponds to the strategy exhibited by the best robots of the best replication performed in the 110 x 110 arena) is characterized by a synchronized target-switching behavior in which the two robots, located in the two different target areas, simultaneously leave their current target area and move directly toward the other target area. The second strategy (that will be called "asymmetrical strategy" from now on and that corresponds to the strategy exhibited by the best robots of the best replication performed in the 150 × 150 arena) is characterized by a switching behavior organized in two phases in which first a robot exits from its target area and travels toward the other target area containing the second robot, and then the latter robot exits from its target area and travels directly toward the target area previously occupied by the former robot.

By testing the robots evolved in simulation in a real environment (i.e., by embodying the neural controller on physical robots and by situating them in the physical environment) we observed that the behaviors exhibited in hardware are qualitatively very similar to those shown in simulation. Examples of the best evolved behaviors both in simulation and in the real environment can be seen at the following webpage: http://laral.istc.cnr.it/esm/evo-communication).

8.3.2 Detailed Analysis of an Exemplar Solution (Asymmetrical Strategy)

In this section we describe in detail the behavioral and communication skills of the best-evolved robots (of the best replication of the experiment performed in the 150 × 150 cm arena) that display an asymmetrical strategy. Moreover, we describe the origin of such skills by analyzing how the behavioral and communication skills exhibited by robots of succeeding generations vary over the course of evolution.

To perform this analysis we divided the overall behavior exhibited by the robots into a list of selected elementary motor and communicative behaviors corresponding to sequences of sensorimotor interactions that produce a given functionality (e.g., that allow a robot to avoid an obstacle, or to move toward the other robot located in the other target area, or to produce a signal that allows the other robot to exit from its current area when the two robots are concurrently located in the two areas). The division of the robots' overall individual behavior into individual elementary behavior has been realized through the use of mutual exclusive conditions (for details see De Greeff and Nolfi 2010). For example, the sequences of robot/environmental interactions in which the robots' infrared sensors are activated above a given threshold are classified as *obstacle-avoidance* behavior (until the infrared sensors are no longer activated). Similarly, the sequences of robot/environmental interactions in which a robot is located on the border of a target area and in which this robot moves forward (and turns slightly left or right) are classified as *follow-border* behavior (provided that the robots' infrared sensors are not activated).

For reasons of clarity, behaviors having similar functions or constituted by sequences of similar but not necessarily identical sensorimotor interactions, or both, are grouped into the same elementary behaviors. For example, sequences of sensorimotor interactions in which the agents produce similar, although not identical, explicit signals are grouped into the same elementary signaling behavior provided that the effect of the signals produced have a similar qualitative effect on the other robot.

Motor and Communication Behaviors Repertoire

In this section we describe the elementary motor and communicative behavior exhibited by the best robots of the last generation. For each elementary behavior we briefly describe the functionality of the behavior (with respect to the task), the conditions in which it is executed, and the actions that are produced during its execution.

• A **signal-A** behavior consists of the emission of a signal in the range [0.9,1.0]. This signal is always produced by robots located outside the black target area that are not detecting obstacles.

• A **signal-B** behavior consists of the emission of a signal in the range [0.0,0.6]. This signal is always produced by robots located in the black target area.

• An **obstacle-avoidance** behavior consists of a sequence of left-turning movements. This behavior is always performed near an obstacle (a wall or another robot) when left, frontal, or right infrared sensors of the robot are activated, regardless of the signals perceived. The robot turns on the spot until the frontal side of the robot is free from obstacles.

• A **move-straight** behavior consists of a sequence of move-forward movements. This behavior is always produced by robots located outside target areas when no other robot is perceived visually and no obstacles are detected.

• A **follow-border** behavior consists of a combination of left-turning and move-forward movements that allow a robot to move counterclockwise by following the border of an area. The *follow-border* behavior is always produced by robots located in the black area that do not visually perceive the other robot, regardless of any perceived signal. This behavior originates evolutionarily from the modification of a *remain-on-black-area* behavior that allows the robot to remain on the target area by producing circular trajectories without necessarily moving along the border independently from whether or not the other robot is visually perceived (see section 8.3.2).

• An **avoid-robot** behavior consists of a sequence of left-turning movements that make the robot turn on the spot until the other robot exits from its field of view. This behavior is produced by robots located outside areas that visually perceive the other robot in all cases, except cases in which additional conditions trigger the execution of the *move-toward-robot* behavior.

• A **move-toward-robot** behavior consists of a sequence of move-forward and left-turning movements that allow a robot to move straight by slightly turning toward the direction of a visually perceived robot. This behavior is always produced by robots that: are located outside target areas, previously visited the white target area, detect signal-B, and detect the other robot in their field of view.

• A **look-robot-and-follow-border** behavior consists of a combination of left-turning, right-turning, and move-forward movements that allow the robot to remain on the border of the area while maintaining the other robot on the left side of its field of view. This behavior also allows the robot to reach a particular location in its target area with respect to the other robot located in the other target area and hence with respect to this other target area. This latter aspect is realized by remaining on the spot when the other visually perceived robot is on the front or right side of the visual field and by moving counterclockwise along the border of the area when the other robot is on the left side of the visual field. This behavior is always produced by robots that are located in the black target area, perceive signal-A, and visually perceive the other robot.

• An **exit-white-area** behavior consists of one or a few move-forward movements that allow a robot located in the white target area to exit from this area. This behavior is always produced by robots located in the white target area that perceive signal-B and visually detect the other robot in the left part of their visual field.

• An **exit-black-area** behavior consists of one or a few move-forward movements that allow a robot located in the black target area to exit from this area. This behavior is always produced by robots located in the black target area that perceive signal-B.

The identification of the robot's elementary behaviors, in this case, is simplified by the fact that this robot displays a reactive behavior, in other words, always reacts in the same way to the same sensory states. For an analysis of other individual solutions in which the internal dynamic occurring within the agents' control system plays a significant role, see De Greeff and Nolfi 2010.

Arbitration and Combination of the Elementary Behaviors

To illustrate how the elementary behaviors previously described are combined and arbitrated to solve the robots' adaptive task we will describe a typical trial (see the videos available from http://laral.istc.cnr.it/esm/evo-communication).

At the beginning of a trial the two robots are located outside target areas. In this phase the robots display a *move-straight* behavior when they are far from obstacles and do not visually perceive other robots, an *obstacle-avoidance* behavior when they detect an obstacle through infrared sensors, and an *avoid-robot* behavior when they visually perceive the other robot. The combination of the *move-straight* and *obstacle-avoidance* behaviors allows the robots to explore the environment. The *avoid-robot* behavior does not play a functional role when both robots are located outside target areas. Indeed, the performance in a normal condition does not significantly differ from the performance in a control condition in which the robots located outside target areas were not allowed to visually detect the other robot. The signaling behaviors produced when both robots are located outside target areas do not alter the motor behavior of the robots themselves and thus do not have any functionality.

When a robot enters the white target area while the other robot is located outside target areas, it starts to produce a *follow-border* behavior. This *follow-border* behavior allows the robot to remain in the white target area until the other robot enters the black target area. The signaling behavior produced by the robot located in the white target area does not have any functionality since it does not alter the motor behavior of the other robot. The implicit signal produced by the robot located in the white target area triggers the *avoid-robot* behavior in the other robot that plays an adaptive role in this circumstance. Indeed, the variance of the overall performance (observed in a normal condition and in a control condition in which the robots located outside target areas were not allowed to visually detect robots located in the white target area) is significant (mean score of 4.723 and 3.941, respectively).

When a robot enters the black target area while the other robot is located outside target areas, it starts to produce a *signal-B* behavior and a *follow-border* behavior or a *look-robot-and-follow-border* behavior, depending on whether or not it perceives the other robot visually. The function of the *follow-border* behavior is to remain in the black target area and to look around in order to identify the relative position of the other robot. The *look-robot-and-follow-border* behavior plays several roles (to be discussed in more detail): (1) it allows the robot to remain in the black target area, (2) it allows the robot to assume a specific position in the target area relative to the other robot that in turn provides for that robot an indication of the exact position of the black target area, and (3) it allows the robot to orient itself toward the center of the white target area (as soon as the other robot enters that target area). Also in this case, the explicit signals produced by the two robots do not affect their motor behavior and therefore do not have any functionality.

Finally, when the two robots are concurrently located in the two target areas they trigger a sequence of coordinated behaviors that is repeated over and over. This allows the two robots to quickly exchange their relative locations several times, thus maximizing their fitness.

During the first phase of this sequence, the robot located in the black target area displays a *follow-border* behavior or a *look-robot-and-follow-border* behavior depending on whether or not it visually perceives the other robot. The robot located in the white target area displays a *follow-border* behavior.

During the second phase, when both robots visually perceive each other on the left side of their field of view, the robot located in the white target area triggers an *exit-white-area* behavior that allows it to exit from the area and to initiate a *move-straight* behavior toward the black target area.

During the third phase the robot that left the white target area displays a *move-toward-robot* behavior, moving toward the other robot while the robot located in the black target area continues to look toward the former approaching robot. The trajectory of the *move-toward-robot* behavior allows the approaching robot to move approximately toward the center of the black target area, thus maximizing the chance to enter this target area and avoiding the risk of obstructing the occupying robot. The *look-robot-and-follow-border behavior*, through which the occupying robot maintains the approaching robot on the left part of its visual field, allows the former robot to leave the black target area while being oriented toward the direction of the white target area.

During the fourth phase, as soon as the approaching robot enters the black target area and switches its signaling behavior from A to B, the occupying robot leaves this target area by triggering an *exit-black-area* behavior and then a *move-straight* behavior. The newly arrived robot triggers a *follow-border* behavior and then a *look-robot-and-follow-border* behavior. The orientation of the robot exiting from the black target area (that depends on the relative position assumed by the robot in this target area, the ability to keep the approaching robot on the left side of its visual field, and the ability of the approaching robot to move toward the center of the area) ensures that the *move-straight* behavior will bring this robot directly toward the center of the white target area.

Finally, during the fifth and last phase, the robot that left the black target area enters the white target area. At this point the two robots are located again in the two target areas and the sequence of coordinated behaviors articulated in the five phases is repeated.

Communication System

In this section we focus on the communication system possessed by evolved robots and on the relation between robots' behavioral and communication skills. More precisely, we will describe the motor behaviors that allow the robots to access the information that has a communicative value, the explicit and implicit signals produced, and the (context-dependent) effect of the detected signals.

The elementary behaviors that allow the robots to access and to generate information that has a communicative value include: an *exploration* behavior (a combination of an *obstacle-avoidance* behavior and a *move-forward* behavior) that allows the robots to identify the location of the two target areas, a *follow-border* behavior that allows the robots to maintain this information over time, and a *look-robot-and-follow-border* behavior that allows the robots to identify and assume a specific position in a target area with respect to the location of the other robot. Interestingly, part of the information conveyed through implicit and explicit signals is not simply extracted from the environment but is instead generated through the behavioral and communicative interaction between the two robots. For example, information that encodes the location of the center of the two target areas (that cannot be detected directly by a single robot) is extracted by the two robots through a coordinated behavior that allows the robots to assume a precise relative position in the target area with respect to the other robot.

The signals produced by the robots include two explicit signals (A and B) that encode whether a robot is located outside or inside the black target area, respectively, and an implicit signal constituted by the body of a robot itself that can be visually detected by the other robot and that provides an indication of its relative position. The fact that the explicit signals do not differentiate the white target area from the regions outside target areas does not constitute a source of ambiguity since this information is exploited only by robots currently located in target areas and because robots never occupy the same target area.

The effects of implicit and explicit signals consist in a modification of the robots' motor behavior that is context dependent (i.e., the type of effect produced or whether or not the effect will be produced, or both, depends on the state of the robot detecting the signal). More precisely:

• the perception of signal-B always triggers an *exit* behavior in robots located in the black target area;
• the perception of signal-B in combination with an implicit signal constituted by the visual perception of the other robot on the left side of the visual field always triggers an *exit* behavior in robots located in the white target area;
• the perception of signal-B in combination with an implicit signal constituted by the visual perception of the other robot triggers a *move-toward-robot* behavior in robots located outside target areas that previously visited the white target area.
• the perception of the implicit signal always triggers an *avoid-robot* behavior in robots located outside target areas (with the exception of the case reported above that triggers the execution of the *move-toward-robot* behavior).

Evolutionary Origin of Robots' Motor and Communicative Skills
The analysis of individuals of successive generations indicates that the behavioral and communication repertoire exhibited by the robots progressively complexifies throughout generations as follows.

1. In the very first generations the robots develop an *exploration* behavior that consists of the combination of *move-forward* and *obstacle-avoidance* behaviors. The exhibition of these behaviors allows the robots to occasionally score 1 point when they happen to transit over the two target areas at the same time.
2. During generations 5–10 the robots develop a *remain-on-black-area* behavior that allows them to remain in the black area when they enter it. The exhibition of this new behavior increases the probability that the two robots happen to be concurrently located in the two areas since it eliminates a situation in which the latter robot enters the white area, while the former robot already has abandoned the black area.
3. The development of a capacity to remain on the black area, however, also has an additional function; it allows the robot located in the black area to access information

that is potentially useful for the other robot when it reaches the white area (information that, as we will see, may allow the other robot to decide whether it should remain or exit from its area). This creates the conditions for the development of an ability to communicate to the other robot whether a robot is located in a black area or not through the production of two different signals (A and B).

4. The development of these *signal-A* and *signal-B* communication behaviors does not lead to an improvement in performance in itself, but creates the adaptive conditions in the next generations for the development of an *exit-black-area* behavior, which is executed by robots located in the black area detecting the signal-B produced by another robot also located in the black area. This new behavior allows the robots to occasionally exchange areas in the following situation: a robot that visited the white area enters the black area that already contains the other robot; through the *exit-black-area* behavior the later robot exits from the black area and subsequently reaches the white area.

5. The development of these behavioral and communication skills, in turn, creates the conditions for the development of a *remain-on-white-area* behavior that allows the robots to remain in the white area until they do not detect the signal that indicates that the other robot is located in the black area. The development of this new behavioral skill eliminates the problem caused by the fact that while the second robot enters the black area the first robot has already exited from the white area.

At this stage of the evolutionary process, the robots are able to reach the two areas through an exploration behavior, to remain on the black area until the other robot also enters the black area. On the basis of these skills, they are able to be located in the two areas for the first time in most of the trials but they are able to switch areas only occasionally. In many cases the trial ends before the robot exiting from the black area succeeds in finding the white area, because, after exiting from the black area, it resumes a simple but time-consuming exploration behavior.

During the next phase of the evolutionary process, however, the robots manage to develop new additional skills that allow them to switch areas more frequently by directly navigating from one area to the other:

6. Around generation 205, the robots develop a *move-toward-robot* behavior that allows the robot exiting from the white area to navigate toward the robot emitting signal-B and therefore directly toward the black area. This new behavioral skill drastically reduces the time needed by the robot located in the white area to reach the black area.

7. Finally, after a long, substantially stable phase, at generation 814, the robots develop a new way to remain in the black area that consists in remaining in the border of the area itself while looking toward the other robot (i.e., the *look-robot-and-follow-border* behavior). Also in this case, this new behavior is realized through the exploitation of the explicit and implicit signals produced by the robots located outside the black area, which in turn are based on the ability to produce the behavioral skills previously

developed. The function of this new behavior is to ensure that the robot located in the black area positions itself toward the other robot located in the white area (or traveling from the white to the black area) and therefore toward the white area. This, in turn, ensures that when the robot exits from the black area, it will travel directly toward the white area (i.e., toward the direction previously occupied by the other robot). The development of this new behavior also creates the adaptive conditions for a further improvement of the *move-toward-robot* behavior previously developed. Indeed, the exhibition of *look-robot-and-follow-border* behavior implies that the robot located in the black area assumes a specific relative position with respect to the other robot located in the white area (i.e., the left side of the area with respect to the other robot). The fact that the robot located in the black area now assumes such a specific position allows the robot traveling from the white to the black area through the *move-toward-robot* behavior to position itself toward the center of the black area, thus minimizing the risk of missing this target. This ability is refined in the following generations.

Overall this analysis shows how the behavioral and communication skills developed by the robots at a certain stage of the adaptive process often create the conditions for the development of further skills with additional functionalities that are based on previously developed skills (figure 8.3). With the sentence "create the conditions for

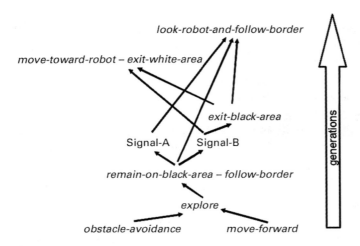

Figure 8.3
Schematic representation of the relations between different behavioral and communication skills. The y-axis indicates the course of the evolutionary process and the order in which skills are developed. The arrows summarize the most important dependencies between the different elementary skills. More precisely, the arrows pointing to a given capacity indicate that the development of the skills at the bottom of the arrow created the adaptive conditions for the development of the new capacity and the fact that the new capacity is based (i.e., depends) on the previously developed skills.

the development of further skills" we mean that the skills that have been developed after would not have been developed (or would have had a lower probability to be developed) without the previously developed skills. With the phrase "are based on previously developed skills" we mean that the newly developed skills require the previously developed skills and that the eventual loss of one or more or the previously developed skills would also imply the loss of the newly developed skills.

The development of new skills that are based on previously developed skills implies that old skills tend to assume additional functions (i.e., support of these newly developed skills). The creation of these chains of dependencies explains why the adaptive processes observed in this and other replications of the experiments can be described fundamentally as an incremental process in which new skills are often developed on top of previously developed skills and in which previously developed skills tend to be preserved in successive generations.

Moreover, the observation that new behavioral and communication skills are often based on simpler previously developed skills implies that the signals that are produced and exploited by the robots are not simply "grounded" on robots' sensorimotor states but also on robots' behaviors.

8.4   Discussion

We believe that the experimental scenario illustrated in this chapter represents a minimalist model that allows us to study how communication can evolve in a population of initially noncommunicating robots and how robots' communication skills can progressively complexify as they adapt to their task/environment. In this section we will discuss how the obtained results can help us to better answer the general questions identified in section 8.1.

The first issue that we want to consider is *under what circumstances and how can communication evolve in the first place*. The evolution of a communication skill, in fact, requires the development of two complementary but interdependent abilities: an ability to produce signals that are useful (from the point of view of the signaler or the receiver, or both) and an ability to react to signals in a way that is useful (from the point of view of the signaler or the receiver, or both). As Maynard Smith puts it: "It's no good making a signal unless it is understood, and a signal will not be understood the first time it is made" (Maynard Smith 1997). From the point of view of the evolution of explicit signaling capabilities, this implies that variations that lead to the production of a useful signal will tend to be retained only if agents already have the complementary ability to react to that signal in an appropriate way. Or, vice versa, variations that lead to an ability to react to signals in a useful way tend to be retained only if agents already have the complementary ability to produce the corresponding signal. This means that adaptive variations that lead to the production of useful signals or to the exploitation of the same signals, but not to both, are adaptively neutral unless the two abilities are developed at the same time. This aspect seems to indicate that the evolution of communication would be an extremely unlikely event, a consideration that is in contrast to experimental evidence.

This apparent paradox can be solved by hypothesizing that: (a) originally neutral traits can later acquire a communicative function, and (b) traits originally playing a certain function can later be exapted (Gould 1977) to play an additional communicative function. This general hypothesis can be further articulated into two cases, depending on whether the preexisting trait consists of the ability to produce an action that could potentially assume a signaling value (as proposed by Konrad Lorenz and other earlier ethologists); or in the tendency to react in a certain way to signals that could potentially assume a communicative value (Maynard Smith and Harper 2003). Evidence supporting the former hypothesis is constituted by the observation that the beak-wiping behavior serving a preening function displayed by several species of grass finches, in some species plays the role of a courtship signal (Morris 1958). Evidence supporting the latter hypothesis is constituted by the colorful phenotype of *Papilio memnon*, which increases the chances of survival of this species by exploiting the tendency of its predator to avoid distasteful insects characterized by a colorful phenotype (Maynard Smith and Harper 2003).

The results obtained through the synthetic experiments presented in this chapter confirm that indeed, communication can emerge despite the fact that the traits that are necessary for its emergence—namely, an ability to produce useful signals and an ability to react to signals appropriately—taken in isolation are adaptively neutral in that they do not, per se, increase the reproductive chances of individuals that possess them. Moreover, the possibility to analyze the course of the evolutionary process in detail, thanks to the synthetic nature of these experiments, allows us to identify how the problem of developing two interdependent traits that are adaptively neutral in isolation is solved. Indeed, the analysis reported in section 8.3.2 indicates that the evolution of communication skills occurs through the exploitation of traits that originally did not serve a communicative function or fulfill any functionality.

An example of a case in which preexisting signal acquires a communication functionality through a variation in the way in which agents react to the signal (and not through a variation of the signal itself) is constituted by the variations occurring from generation 210 on that lead to the development of the move-toward-robot behavior (section 8.3.2). Up to this point the signal-B, which is produced by robots located in the black area, triggers an *exit* behavior in the robot located in the white area (thus allowing the robots to eventually reach the other area). From generation 210 on, the same signal plays an additional functionality that is realized by triggering a move-toward-robot behavior in robots that previously visited the white area. The new functionality is achieved through a variation that modifies the way in which the robots react to the signal, but not the signal itself or the conditions in which it is produced.

An example of a case in which a preexisting ability to react to signals in a specific way acquires a functionality through a variation of the signal produced, but not of the way in which the robots react to the signal, is constituted by the development of the signal-B that triggers the *exit-black-area* behavior. The tendency to react to this signal by exiting from the black area, in fact, is displayed already from generation 10 on (which was be observed in a control condition in which one robot is forced by the experimenter to produce the signal-B while the other robot is located in the black area). The ability to produce signal-B in the black area is developed several generations after. The presence of a trait that makes the robot exit from the black area in reaction to signal-B is thus exploited by developing an ability to produce the signal-B in a situation in which the *exit-black-area* behavior is functional.

The second issue that we want to consider is *how and to what extent the evolved communication system can complexify*. Complexity can be measured along different dimensions. One dimension concerns the number of different elementary behaviors produced by the agents. A second dimension concerns the number of signals or combination of signals serving a communicative function that co-determine the expressive power of the communication system. A third dimension concerns the diversification of the effects that each signal produces depending on the context in which the signal is detected. A fourth dimension concerns the ability of the agents to access and to generate information that has a communicative value and that can then be conveyed through communication signals. Finally, a fifth dimension concerns the nature of signals developed, namely whether a signal encodes information directly available through the agents' sensors or more complex, re-elaborated information (Mirolli and Nolfi 2010).

The analysis of the evolutionary process as described indicates that improvements in terms of performance are often correlated with a complexification of agents' skills with respect to one or more of these five dimensions. The comparative analysis of different replications of the experiments also shows how solutions that are comparable in terms of performance and in terms of overall complexity of the evolved strategy can differ significantly with respect to the complexity along different dimensions.

In the case of the two best replications of the experiments performed in large and small arenas, evolved individuals display a rich behavioral and communicative repertoire that includes seven to ten different elementary behaviors and four to six signals (constituted by different explicit signals or combination of implicit and explicit signals), each producing one to three different effects depending on the context in which signals (or combination of signals) are experienced.

With respect to the ability of the robots to access, generate, and elaborate communicative information, in most of the cases explicit signals encode, on the one hand, nonabstract information that is directly and currently available through the sensors of the robots and that is accessed through the exhibition of simple behaviors (i.e., an exploration or a remain-on-target behavior, or both). Nonabstract signals of this form do not involve a significant re-elaboration of the sensory information or the integration of sensorimotor information through time, or both (Hauser 1996; Rendall et al. 1999). In the case of the symmetrical strategy, however, explicit signals encoding abstract information are also observed (see the analysis reported at http://laral.istc.cnr .it/esm/evo-communication).

Implicit signals and combination of implicit and explicit signals, on the other hand, often encode abstract information. This can be explained by considering that the implicit signal constituted by the actual position of a robot often implicitly encodes useful information concerning the sensory states and motor actions previously perceived and performed by the robot itself. The need to extract and communicate information about previous experienced sensory states therefore is solved by selecting behavioral skills that allow the robots to integrate and elaborate information by acting in the environment rather than by performing internal operations. An example of an abstract signal is constituted by the combination of the implicit signal and the explicit signal-B produced by a robot located in the black area that allows the other robot to infer the direction in which it should navigate to reach the central part of the black area—information that is not directly available from the state of the robot's sensors and that reflects the effects of the previous sensorimotor interactions between the robot and the environment (see section 8.3.2).

All used signals are deictic (i.e., they provide information that is dependent on the current context of the sender (Hockett 1960)). Displaced signals (i.e., signals providing information that is independent from the current context of the sender (Hockett 1960)) are not observed. Finally, most of the used signals are informative/ manipulative (i.e., they convey information possessed by one of the individuals, or one individual manipulates a second individual to accomplish a certain adaptive function). In few cases, however, relational signals are also observed. By relational signals we mean signals that are generated through a communicative interaction, allowing a group of individuals to perform a collaborative task that could not be accomplished by a single individual (i.e., signals analogous to the vocal duetting produced by several species that allow them to establish and maintain a pair bond, Bailey 2003; Farabaugh 1982; Haimoff 1986). An example of relational signal is observed in the experiment displaying the symmetrical strategy. The signaling interaction occurring between the two robots allows the robots to produce two different signals; one occurs when only one robot is located outside target areas, the other when both robots are located outside target areas (see De Greeff and Nolfi 2010). On the evolution of relational signals in a similar experimental setting see also Marocco and Nolfi 2007.

The third issue that we want to consider is *the role of innovations in the evolutionary process and the relation between social/communicative interaction and open-ended evolution.* The analysis of the origins of robots' behavioral and communicative skills demonstrates, on the one hand, how the evolutionary process is strongly influenced by the initial capabilities of the robots. These, in turn, depend on the family of strategies on which the evolutionary process converges in the very initial generations as a result of the random generation of the genome of the initial population, and as a result of the high stochasticity that characterizes the selection process in the very first generations. On the other hand, the analysis of the most successful replications of the experiment also demonstrates how robots' skills can be progressively transformed and how the number and the complexity of the robots' elementary skills can increase during the adaptive process until optimal or close to optimal solutions are discovered. As we mentioned in section 8.3.2, such progressive complexification of robots' skills seems to occur as a result of an incremental process in which the development of new skills often creates the adaptive condition for the development of further skills and in which previously developed skills tend to be retained.

The tendency to preserve previously developed skills can be explained by considering that new skills often exploit (are based on) previously developed skills. For example, in the case of the experiment displaying the asymmetrical strategy described earlier, the *move-toward-robot* behavior that allows the robots located in the white area to navigate directly toward the black area depends on the *follow-border* behavior exhibited by robots remaining on the black area that was initially developed to allow the robots to reach the two areas for the first time, not for switching areas. Moreover, the *move-toward-robot* behavior also depends on the *signal-B* behavior that was previously developed to allow one robot to exit from the black area when the other robot is also located in the same area. This means that the *move-toward-robot* behavior is based on the other two previously developed behavioral skills and that the development of the *move-toward-robot* behavior causes the *follow-border* and *signal-B* behaviors to acquire additional functionality—that of supporting the *move-toward-robot* behavior.

The ability to generate the required new skills can be explained by considering the potential to exploit previously developed skills. The development of new skills, in fact, not only leads to an improvement of agents' performance but also often leads to the establishment of the adaptive condition, which enables the development of further and more complex skills. For example, the ability to remain on the black area by displaying a *look-robot-and-follow-border* behavior (i.e., by assuming a precise position in the target area with respect to the other robot) creates the condition for the development of an ability to leave the white area by navigating toward the center of the black area. More generally, concerning communication and social interaction, the development of an ability to signal relevant information enriches the perceptual environment of the robots, thus creating the adaptive conditions for the emergence of new skills which exploit information encoded in detected signals. In other words, innovations often creates the adaptive conditions for the development of additional novelties, thus producing an evolutionary process that is open-ended (within the limits imposed by the complexity of the task/scenario).

Supplemental Data

For supplemental data including movies of the behaviors displayed by evolved robots of different replications of the experiment, go to http://laral.istc.cnr.it/esm/evo -communication. Open software for replicating the experiments in simulation as well as hardware including the source codes, a manual, a tutorial, and the sample files of the experiment can be downloaded from http://laral.istc.cnr.it/evorobotstar/.

References

Bailey, W. J. 2003. Insect duets: Underlying mechanisms and their evolution. *Physiological Entomology* 28:157–174.

Baldassarre, G., S. Nolfi, and D. Parisi. 2003. Evolving mobile robots able to display collective behaviour. *Artificial Life* 9:255–267.

Beer, R. D. 2003. The dynamics of active categorical perception in an evolved model agent (with commentary and response). *Adaptive Behavior* 11 (4): 209–243.

Cangelosi, A., and D. Parisi. 2002. Computer simulation: A new scientific approach to the study of language evolution. In *Simulating the Evolution of Language*, ed. Angelo Cangelosi and Domenico Parisi, 3–28. London: Springer-Verlag.

De Greeff, J., and S. Nolfi. 2010. Evolution of implicit and explicit communication in mobile robots. In *Evolution of Communication and Language in Embodied Agents*, ed. S. Nolfi and M. Mirolli, 179–214. Berlin: Springer-Verlag.

Di Paolo, E. A. 1997. An investigation into the evolution of communication. *Adaptive Behavior* 6 (2): 285–324.

Di Paolo, E. A. 2000. Behavioral coordination, structural congruence and entrainment in a simulation of acoustically coupled agents. *Adaptive Behavior* 8 (1): 25–46.

Farabaugh, S. M. 1982. The ecological and social significance of duetting. In *Acoustic Communication in Birds*, ed. D. E. Kroodsma and E. H. Miller, 85–124. New York: Academic Press.

Floreano, D., S. Mitri, S. Magnenat, and L. Keller. 2007. Evolutionary conditions for the emergence of communication in robots. *Current Biology* 17:514–519.

Gould, S. J. 1977. *Ontogeny and Phylogeny*. Cambridge, MA: Harvard University Press.

Haimoff, E. H. 1986. Convergence in the duetting of monogamous old world primates. *Journal of Human Evolution* 15:767–782.

Harnad, S. 1990. The symbol grounding problem. *Physica D. Nonlinear Phenomena* 42:335–346.

Hauser, M. D. 1996. *The Evolution of Communication*. Cambridge, MA: Bradford Books/MIT Press.

Hockett, C. F. 1960. The origin of speech. *Scientific American* 203:88–96.

Kirby, S. 2002. Natural language from artificial life. *Artificial Life* 8:185–205.

Marocco, D., and S. Nolfi. 2007. Emergence of communication in embodied agents evolved for the ability to solve a collective navigation problem. *Connection Science* 19 (1): 53–74.

Maynard Smith, J. 1993. *The Theory of Evolution*. Cambridge, UK: Cambridge University Press.

Maynard Smith, J., and D. Harper. 2003. *Animal Signals*. Oxford, UK: Oxford University Press.

Mirolli, M., and S. Nolfi. 2010. Evolving communication in embodied agents: Theory, methods, and evaluation. In *Evolution of Communication and Language in Embodied Agents*, ed. S. Nolfi and M. Mirolli, 215–220, Berlin: Springer-Verlag.

Mondada, F., and M. Bonani. 2007. The e-puck education robot. http://www.e-puck.org/ (accessed June 11, 2003).

Morris, D. 1958. The comparative ethology of grass-finches (Erythrurae) and mannikins (Amadinae). *Proceedings of the Zoological Society of London* 131:389–439.

Nolfi, S. 2005. Emergence of communication in embodied agents: co-adapting communicative and non-communicative behaviours. *Connection Science* 17 (3–4): 231–248.

Nolfi, S. 2009. Behavior and cognition as a complex adaptive system: Insights from robotic experiments. In *Philosophy of Complex Systems, Handbook on Foundational/Philosophical Issues for Complex Systems in Science*, ed. C. Hooker, Part IV, 443–463. Amsterdam: Elsevier.

Nolfi, S., and D. Floreano. 2000. *Evolutionary Robotics: The Biology, Intelligence, and Technology of Self-Organizing Machines*. Cambridge, MA: Bradford Books/MIT Press.

Nolfi, S., and D. Marocco. 2001. Evolving robots able to integrate sensory-motor information over time. *Theory in Biosciences* 120:287–310.

Nolfi, S., and M. Mirolli. 2010. *Evolution of Communication and Language in Embodied Agents*. Berlin: Springer-Verlag.

Quinn, M. 2001. Evolving communication without dedicated communication channels. In *Advances in Artificial Life: Sixth European Conference on Artificial Life (ECAL 2001)*, ed. J. Kelemen and P. Sosik, 357–366. Berlin: Springer-Verlag.

Quinn, M., L. Smith, G. Mayley, and P. Husbands. 2003. Evolving controllers for a homogeneous system of physical robots: Structured cooperation with minimal sensors. *Philosophical Transactions of the Royal Society of London, Series A: Mathematical, Physical and Engineering Sciences* 361:2321–2344.

Rendall, D., D. L. Cheney, R. M. Seyfarth, and M. J. Owren. 1999. The meaning and function of grunt variants in baboons. *Animal Behaviour* 57:583–592.

Steels, L. 2003. Evolving grounded communication for robots. *Trends in Cognitive Sciences* 7 (7): 308–312.

Trianni, V., and M. Dorigo. 2006. Self-organisation and communication in groups of simulated and physical robots. *Biological Cybernetics* 95:213–231.

Wagner, K., J. A. Reggia, J. Uriagereka, and G. S. Wilkinson. 2003. Progress in the simulation of emergent communication and language. *Adaptive Behavior* 11 (1): 37–69.

Williams, P. L., R. D. Beer, and M. Gasser. 2008. Evolving referential communication in embodied dynamical agents. In *Artificial Life XI: Proceedings of the Eleventh International Conference on the Simulation and Synthesis of Living Systems*, ed. S. Bullock, J. Noble, R. Watson, and M. A. Bedau, 702–709. Cambridge, MA: MIT Press.

9 Evolving Cooperation: From Biology to Engineering

Sabine Hauert, Sara Mitri, Laurent Keller, and Dario Floreano

9.1 Introduction

Robots are increasingly being used to solve real-world tasks such as vacuuming or assembly-line work in industrial applications. Controllers for these robots are typically designed by engineers following textbook guidelines. Although this methodology has proven to be very successful in such applications, it quickly meets its limitations as tasks become more complex. Collective robotic systems, where groups of robots cooperate to solve a distributed task in partially unknown environments, are an example of systems that are difficult to engineer following a classical approach (Beni 2004; Sahin 2005). This is because it is not obvious how to design controllers for individual robots that cooperate toward a common goal.

Evolutionary robotics (ER) has proven to be highly successful in solving difficult or underdefined engineering problems due to its potential to automatically find simple and efficient solutions (Cliff, Husbands, and Harvey 1993; Nolfi and Floreano 2000). However, for the approach to reach its full potential in solving real-world problems, we believe that a better understanding of the influence of different factors driving evolution should be developed and summarized as guidelines. An additional step is then needed to practically use the evolved controllers in a verifiable and adaptable manner.

As a starting point in constructing guidelines for the evolution of cooperative robots, we turn to the biological systems that inspired evolutionary robotics. Over billions of years, animals have evolved to solve a variety of collective tasks from navigation to collective hunting, which evolutionary biologists have studied extensively. By tapping into decades of research in biology, we explore whether the insights concerning the conditions that allow for the evolution of cooperation in nature can be translated into evolutionary algorithms that are applicable to robotic problems. For this purpose, we test the biological predictions on a preliminary robotic experiment. The results obtained from this initial study are then used as a guideline for solving a problem where we evolve a group of flying robots in simulation that must cooperate in forming and maintaining an aerial communication network in a rescue scenario. The simple, efficient, yet unintuitive solutions discovered through this evolutionary process are then reverse-engineered and implemented in hand-designed controllers. This approach is practical for real-world applications because hand-designed controllers can be easier to understand and to parameterize for different scenarios than evolved controllers.

9.2   Understanding the Evolution of Cooperative Behavior

9.2.1   Cooperative Behavior in Animals

Cooperative behavior has constituted one of the biggest mysteries in evolutionary biology, and perhaps in modern biology as a whole (Dugatkin 2002; Lehmann and Keller 2006; Sachs et al. 2004; West et al. 2007). This is because the theory of Darwinian selection predicts that individuals should maximize their own reproduction, rather than altruistically increasing the reproductive success of others. However, evolutionary biology has come a long way in understanding cooperation by determining two mechanisms that may lead to the evolution of cooperative behaviors in groups of conspecifics. First, high relatedness between individuals in a group is expected to promote cooperative behavior within the group. This theory was formalized by Hamilton (1964) and is thus commonly referred to as "Hamilton's rule," "inclusive fitness theory," or "kin selection theory." A second theory that provides an explanation for the evolution of cooperative behavior is that of "group selection" (Dugatkin and Reeve 1994; Lehmann and Keller 2006; West, Griffin, and Gardner 2007). This theory states that selection *between* groups of individuals should result in cooperation *within* groups, regardless of within-group relatedness. It has recently been shown that the two theories are mathematically equivalent (Dugatkin and Reeve 1994; Hamilton 1975; Lehmann et al. 2007).

9.2.2   Cooperative Behavior in Robots

When engineering collective robotic systems, we are interested in maximizing performance. In many collective robotic tasks, performance can be increased if robots in a group cooperate toward a common goal. However, it remains unclear how groups of robots should be composed and selected to achieve maximal performance. In fact, a variety of methods are used in studies that report on evolving cooperative behavior in groups of robots (see Waibel, Keller, and Floreano 2009 for a review), yet few explicitly motivate their choice of evolutionary parameters.

   To study how cooperative behavior can evolve, we designed an experimental setup consisting of groups of robots that could emit and perceive light and were evolved to solve a foraging task. Because cooperative communication can potentially increase the performance of robot groups, this system allowed us to explore whether high

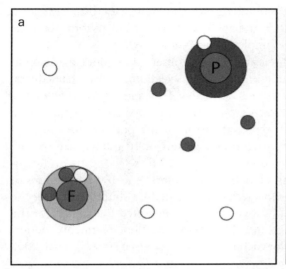 

Figure 9.1

Experimental setup. (a) A food and poison source, both emitting red light, are placed 1 m from one of two opposite corners of the square (3m × 3m) arena. Robots (small circles) can distinguish the two by sensing the color of the circles of paper placed under each source using their floor sensors when driving over the paper. (b) The robot used for the experiments is equipped with two tracks to drive, an omni-directional (360 degree) vision camera, a ring of lights used to emit blue light, and floor sensors to distinguish food and poison sources.

relatedness and group-level selection can lead to an increase in cooperative behavior, as biological theory predicts, and thus an increase in group performance.

In our experimental system, ten s-bot robots (Mondada et al. 2004) placed randomly in a square arena must find a food source emitting red light and avoid a similar red poison source (figure 9.1a). The sources could only be distinguished once the robots were very close to them, by using their floor sensors to detect a disc of colored paper placed under each of the sources. Robots could cooperate by emitting blue lights in a way that provided information on the location of the food and poison sources to other robots. Each robot was controlled using a feedforward neural network, which processed blue and red light, in addition to the information on its own location (at food, poison, or elsewhere) to determine the speeds of its two tracks and whether to emit blue light or not. The weights of the neural network formed the genome of the robot, which evolved over 500 generations in a population of 1,000 robots. The performance was calculated for each robot by counting the number of time-steps within the sixty-second trial during which it was at the food minus the number of time-steps spent by the poison, averaged over ten consecutive trials. At each generation the 200 best-performing robots were selected, replicated through cloning or crossover (with a probability of 0.2), and mutated (with a probability of 0.01 per bit) to form the next generation. For more details, see for example Floreano et al. 2007.

To test the effect of varying relatedness and the level of selection, four experimental treatments were used: (1) high relatedness, group-level selection; (2) high relatedness, individual-level selection; (3) low relatedness, group-level selection; and (4) low relatedness, individual-level selection. To form groups of low-relatedness individuals, we randomly selected 1,000 robots (with replacement) from the pool of the 200 best-performing robots and assigned them to 100 new groups of 10 robots each for the next generation. In contrast, high relatedness was achieved by selecting 100 individuals from the pool of 200, and cloning each 10 times to form the 100 new groups of 10 identical robots. Group-level selection was implemented by simply assigning the same performance score to all robots in a group that represented the average of the individual scores. Alternatively, in the individual-level selection treatments, performance was calculated independently for each individual robot (see Floreano et al. 2007 for details).

In the two treatments where relatedness between robots was high, performance was significantly higher than when relatedness was low (MannWhitney test, all $P < 0.001$, figure 9.2). Group-level selection also resulted in higher performance when robots were highly related ($P < 0.05$). However, when relatedness was low, robots selected at the group level performed significantly worse than those selected at the individual level ($P < 0.001$).

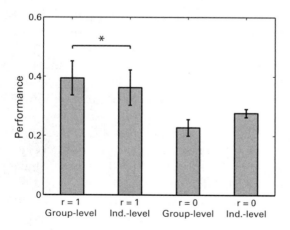

Figure 9.2
Mean (±SD) performance of robot populations during the last fifty generations for each of the four treatments (twenty replicates per treatment). "*" indicates that the bars are significantly different at $P < 0.05$. Bars that are not compared are significantly different at $P < 0.001$.

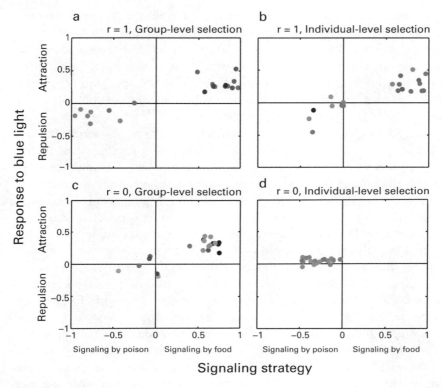

Figure 9.3

Relationship between signaling strategies and responses to blue light in the four treatments. Each dot is the average for the 100 colonies in one replicate after 500 generations of selection. Positive values for the signaling strategy indicate a tendency to signal close to the food, and negative values indicate a tendency to signal close to the poison. Positive values for the responses to blue light indicate attraction to blue light and negative values indicate repulsion (see Floreano et al. 2007 for definitions). The darkness of the points is proportional to the mean performance.

Although this performance comparison seems to indicate that unrelated robots selected at the group level did not cooperate, an analysis of their behavior shows otherwise. Both high relatedness and group-level selection led to the evolution of altruistic communication (figure 9.3a, b, and c). For these treatments, two different communication strategies were observed. In some evolutionary runs, robots produced blue light in the vicinity of the food and were attracted to blue light, thus being likely to end up by the food (e.g., top right quadrant in figure 9.3a). In other runs, blue light was more likely to be emitted by the poison, and resulted in robots driving in the opposite direction and avoiding poison (e.g., bottom left quadrant in figure 9.3a).

These two strategies, although not equally efficient, are both of a cooperative nature, because signalers and receivers evolved complementary strategies. However, when robots were unrelated and selected at the level of the individual, they emitted light by the poison more often than by the food, but were nevertheless slightly attracted to blue light (figure 9.3d). This strategy essentially constitutes a suppression of information, and can therefore be seen as uncooperative behavior (for more information on this strategy, see Mitri, Floreano, and Keller 2009).

These results are interesting in two respects. On the one hand, they show that the predictions of evolutionary theory hold in the case of evolutionary robotics. Since it is difficult to conduct similar experiments in living organisms, this test, within its limitations, provides some supporting contribution to theoretical discussions in biology (see Floreano et al. 2007 and Mitri, Floreano, and Keller 2009 for a discussion on this perspective). On the other hand, from an engineering perspective, it is interesting that cooperative behavior between robots does not always imply high performance and that this depends on the selection method. By designing an evolutionary algorithm using unrelated robots and selecting them at the group level, our results confirm that cooperative behavior between the robots can evolve, as expected from theoretical predictions. However, their performance has been found to be low compared to other selection methods, such as selecting related individuals at the group level. This is due to the inherent inefficiency of this particular selection algorithm. Because robots are all different in a group and the group is selected as a whole, it is difficult to select high-performing individuals, while simultaneously avoiding selecting individuals with low performance (see also "credit assignment problem" described in Waibel, Keller, and Floreano 2009). In addition, the composition of the groups changes at every generation, making it difficult for individuals to optimize their behavior with respect to the behavior of their group mates. Similar results have been obtained in a systematic study of collective object transportation by Waibel, Keller, and Floreano (2009).

In summary, we find that cooperation can evolve either if robots in a group are clones or if they share their performance scores with other members of their group. The highest performance in cooperative tasks is achieved when both these conditions are true.

9.3   From Biology to Engineering

The experiments described in the previous section allowed us to test biologically motivated theories on the evolution of cooperation, and at the same time, to draw guidelines on the design of evolutionary algorithms for groups of cooperating robots. In this section we show how these guidelines can be applied to engineer a solution to a real-world problem. Engineering consists in finding the best possible solution to a

Figure 9.4
Artistic view of the use of a group of flying robots to establish communication networks between rescuers on the ground in a flood scenario.

problem under some constraints. In the context of this chapter, this amounts to selecting the most appropriate evolutionary algorithm that is likely to lead to cooperation between robots and high performance. However, regardless of the performance of the evolved solution, it is often difficult to apply it directly to a real-world problem, because it is likely to be less predictable than a hand-engineered solution, which may result in costly failures. Alternatively, a process of reverse engineering may be applied to the evolved solution to derive a controller whose behavior and operating conditions are predictable while still capturing the simplicity and efficiency of the evolved solution.

To illustrate this process, we describe a situation where groups of flying robots must form communication networks between two rescuers in a disaster scenario as shown in figure 9.4. The robots are fully autonomous, such that the network of robots can be deployed by a single nonexpert rescuer on the ground. To create and maintain wireless bridges and avoid getting lost, the robots must distribute to find rescuers on the ground while staying within the communication range of one another. Flying robots have the advantage of navigating above obstacles while providing unobstructed wireless transmissions. The robots are required to work in environments with no access to GPS satellites or visual information (urban canyons, occluded environments, night operation). Therefore, they do not know their own position or the position of their neighbors. Instead they use local wireless communication and have proprioceptive sensors such as a compass, an altitude, and a speed sensor.

This problem is challenging because existing controllers for flying robots rely on position information and because there is no obvious strategy to design individual controllers that will lead to an effective communication network (Hauert, Zuffery, and Floreano 2009a). Furthermore, the performance of the robots can be measured only at the level of the team as a function of the quality of the resulting communication among rescuers.

9.3.1 Evolving a Group of Flying Robots

To explore this problem, we consider a simplified scenario in simulation, in which a group of robots must deploy and maintain a wireless communication network between two rescuers on the ground (Hauert, Zuffery, and Floreano 2009a). Twenty robots are launched by one rescuer at a rate of 1 every 15 s (±7.5) and the group must then cooperate to find a second rescuer positioned within a ±30-degree angle of a predefined search direction and a distance of 500 ± 50 m. Once the communication link between the two rescuers is established, it must be maintained until the end of the mission, which lasts a maximum of thirty minutes. The robots are simulated using a physics engine in which we implement a first-order dynamics model of a fixed-wing robot that flies at a speed of 10 m/s and turns with a minimum turn radius of 20 m. These constraints bring interesting dynamics to the system since the robots cannot stop or turn on the spot like ground robots or hovercrafts. The communication range of robots and rescuers is of maximum 100 m with added noise between 90 m and 100 m.

Each robot is controlled using a feedforward neural controller consisting of three inputs, four hidden neurons, and one output controlling the turn rate of the robot (speed and altitude are constant). The first input to the network is the heading of the robot given by a magnetic compass. The second and third inputs are the number of network hops separating the robot from the two rescuers (high values indicate that the robot is disconnected), where network hops can be seen as the number of times a message sent from a rescuer needs to be forwarded from one robot to another before it reaches the robot in question (number of lines between a rescuer on the ground and a robot in figure 9.4). The genome of each robot consists of the 16 synaptic weights of the neural network, each represented by 8 bits, making a total genome size of 128 bits.

Based on the results obtained in the previous section, we use homogeneous groups and apply group-level selection. The performance of each group is computed as the minimum number of robots that need to fail for the communication between the rescuers to break, averaged over thirty minutes and ten missions. This performance measure favors the rapid creation of communication pathways and the robustness of the network over time. As in the foraging experiment with homogeneous individuals, a population of 100 genomes is used, which are cloned twenty times to construct 100

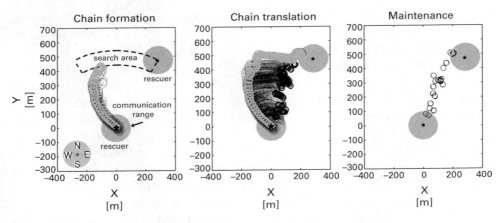

Figure 9.5
Trajectories of the robots with the best evolved controller (over all populations in all generations) during a thirty-minute mission. In this mission, the second rescuer is to the northeast of the launching rescuer to show the full extent of the chain translation displayed by the group. The trajectory of the first launched robot is shown by a light grey line.

groups of twenty robots each. After ranking the genomes according to the measured performance of the robot groups, the twenty best genomes in the population are copied to the new population (elitism) and cloned to make groups of twenty robots each. The remaining population is generated by repeatedly selecting two random individuals from the best 30 percent of the genomes, applying one-point crossover to the pair with a probability of 0.2 and then mutating the newly created individual with a probability of 0.01 per bit, and cloning it twenty times.

With these settings, artificial evolution results in controllers for flying robots that are able to create and maintain a communication network between the rescuers. At the end of the evolutionary process, robots with the best evolved controller over all populations in all generations were tested in 1,000 consecutive missions, of which 975 missions led to the creation and maintenance of a network between the two rescuers. As shown in figure 9.5, these robots form a chain that translates over the area to be searched until the second rescuer is found. The chain then stays on the spot to maintain the communication link.

9.3.2 Reverse Engineering
Robot controllers for real-world applications must often adapt across different scenarios depending on the needs of a given operation (different environment, different number of robots, slightly different task, etc.). However, evolved controllers are constrained to scenarios for which they were evolved. Furthermore, evolved controllers

such as neural networks, electronic circuits, and programs are not always easy to understand. This makes it difficult to evolve robots that are rapidly and robustly usable out of the box in unexpected situations. Possible solutions to this challenge include evolving a different controller before each operation. For this to be practical, the evolutionary process must be extremely rapid and portable. While most current evolutionary experiments are conducted over several hours or even weeks on large computer clusters, the natural increase in computational resources and power might allow for such an approach in the future. Moreover, one could imagine evolving a controller that takes as an input the parameters of the environment. This is indeed a promising approach, although it is currently challenging to find optimal controllers for different combinations of parameters because of current limitations in evolving multi-objective systems (Urzelai and Floreano 2001). Another solution would be to allow the system to evolve online, provided that it can be given some time to fail and learn (Floreano and Mattiussi 2008). This is not necessarily obvious for all applications, including search-and rescue missions.

Here, we propose to address this issue by reverse engineering high-performing controllers found by the evolutionary algorithm. In doing so, we aim to build a control model with a limited set of variables that captures the simplicity and efficiency of the evolved solution. To proceed, we analyze the effect of each input of the best evolved neural controller on the turn rate of the robot (see Hauert, Zuffery, and Floreano 2009b for details). Through this systematic analysis, we identify three simple behaviors performed by the individual robots:

1. Robots that are connected to the launching rescuer, even indirectly, move away from it (figure 9.6, low hop values).
2. Robots that are disconnected from the launching rescuer, move toward it with a different radius than when connected (figure 9.6, high hop values).
3. Robots connected to both rescuers turn following small circular trajectories.

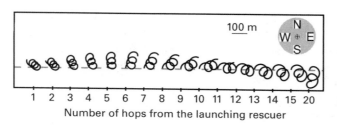

Figure 9.6
Effect of the number of hops that separate the launching rescuer from a robot on its trajectory. Here, we plot the trajectories of the best evolved controller over 30 s. The robot was never connected to the second rescuer during these experiments.

The effect of these individual behaviors on the behavior of the entire group can be hypothesized as follows. As long as robots are being launched, they remain connected to the launching rescuer (at least indirectly) while advancing in a common direction. This results in the formation of a chain. Once all robots have been launched, the chain continues to advance until it disconnects from the launching rescuer. To reconnect, the chain changes direction and moves toward the launching rescuer. Not only does the chain reconnect, but it also translates along the communication range of the rescuer, effectively sweeping through the search area and eventually finding the second rescuer. This is due to the different turn radius of the robots when disconnected or connected to the launching rescuer. Finally, robots connected to both rescuers maintain the communication pathway by performing small circular trajectories.

To explore whether the extracted behaviors yield similar collective behavior as the evolved controllers, we then translate the rules into an algorithm (Hauert, Zuffery, and Floreano 2009b). To do so, we simplify the strategy found through evolution by only considering if a robot is receiving messages from the rescuers (i.e., whether it is connected to the rescuers rather the number of hops separating it from the rescuers). The robot is assumed to fly at a constant speed v and to form circles of radius r_{min} or more. The reverse-engineered controller, summarized as follows, outputs the turn rate ω of the robot based on the global direction ψ_s in which the robots must search for the second rescuer, the orientation of the robot ψ, and whether the robot is connected or disconnected from the rescuers.

When the robot is connected to the launching rescuer only (figure 9.7a):

$$\omega = \begin{cases} \dfrac{v}{r_1} & \text{if obtuse}(\psi, \psi_s) \\ \dfrac{v}{r_2} & \text{otherwise} \end{cases}$$

When the robot is disconnected from the launching rescuer (figure 9.7b):

$$\omega = \begin{cases} \dfrac{v}{r_3} & \text{if obtuse}(\psi, \psi_s + \pi) \\ \dfrac{v}{r_4} & \text{otherwise} \end{cases}$$

Finally, when the robot is connected to both rescuers (figure 9.7c):

$$\omega = \begin{cases} \dfrac{v}{r_{min}} \end{cases}$$

where obtuse returns true if there is an obtuse angle between the two variables and r1, r2, r3, and r4 are parameters of the controller.

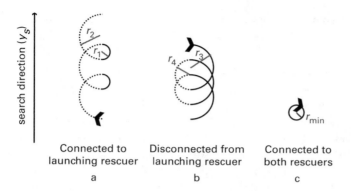

Figure 9.7
Robot trajectories performed by a reverse-engineered controller with parameters r_1, r_2, r_3, and r_4 depending on the connection to the rescuers and the general search direction ψ_s.

Using these rules, we are able to reproduce the strategies found through evolution, namely chain formation, translation, and network maintenance. Furthermore, the rules are easier to understand than a neural network and the trajectories of the robots can be mathematically modeled and subsequently parameterized (by setting r_1, r_2, r_3, and r_4) for a different desired area coverage. This can be intuitively explained by the fact that the controller is based on simple geometry (circular trajectories). Therefore, changing the ratio between parameters r_1 and r_2 will lead to robot trajectories that move away from the launching rescuer at different speeds and thus affect the length of the robot chain. Furthermore, the relationship between the trajectory of a robot when connected to the launching rescuer (defined by r_1, r_2) or disconnected (defined by r_3, r_4) will determine the speed at which the chain translates.

9.4 Conclusion

In this chapter, we have taken inspiration from the predictions of evolutionary biology regarding the evolution of cooperative behavior, and systematically compared the effect of different evolutionary parameters on a collective system of foraging robots evolved artificially. Through these experiments, we have derived a set of guidelines, which state that cooperative behavior, as well as a high group performance can be achieved if groups are composed of genetically identical individuals and selected at the level of the group. We then applied these guidelines to a real-world problem where flying robots with limited sensory capabilities were required to create communication networks in disaster areas. Finally, we proposed to reverse engineer the best evolved solution to design a robot controller model whose behavior is well understood and predictable.

This chapter presents the first step toward evolving groups of robots for real-world problems. Although the results shown here have been conducted in simulation, the reverse-engineered controllers have also been implemented on board real flying robots (Hauert 2010; Hauert et al. 2013). In addition, we aim toward more complex scenarios involving windy environments, increasing the number of rescuers or allowing for mobile rescuers. Our approach has thus shown the potential of evolutionary robotics in generating efficient control solutions to complex engineering problems, such as controlling groups of robots.

More generally, we believe that the biological literature is a promising starting point to understanding many aspects of evolutionary systems. This is because many of the factors influencing systems of evolving robots have been extensively studied by biologists. The results we have reported in this chapter are only the tip of the iceberg, however. Many open questions remain on how to design evolutionary robotic systems and how to apply them to complex real-world applications. In particular, there is still a widely untapped source of biological theories that could be used for the purpose of ER, such as theories concerning division of labor, coevolution, and neuroscience. These guidelines extracted from biology to evolve controllers for robots may potentially lead to the compilation of a complete "reference manual" on how to evolve robots that can solve competitive tasks outside the lab.

Acknowledgments

This work is supported by armasuisse, competence sector Science + Technology for the Swiss Federal Department of Defense, Civil Protection and Sports; Swiss NSF grant no. K-23K0117914/1 on the "Evolution of Altruistic Communication"; and the "Swarmanoid Project," a Future Emerging Technologies (FET IST-022888) project funded by the European Commission.

References

Beni, Gerardo. 2004. *From Swarm Intelligence to Swarm Robotics. Swarm Robotics.* Lecture Notes in Computer Science. Berlin, Heidelberg: Springer.

Cliff, Dave, Phil Husbands, and Inman Harvey. 1993. Explorations in evolutionary robotics. *Adaptive Behavior* 2:73–110.

Dugatkin, Lee Alan. 2002. Cooperation in animals: An evolutionary overview. *Biology and Philosophy* 17:459–476.

Dugatkin, Lee Alan, and Hudson Kern Reeve. 1994. Behavioral ecology and levels of selection: Dissolving the group selection controversy. *Advances in the Study of Behavior* 23:101–133.

Floreano, Dario, and Cladio Mattiussi. 2008. *Bio-Inspired Artificial Intelligence: Theories, Methods, and Technologies*. Cambridge, MA: MIT Press.

Floreano, Dario, Sara Mitri, Stephane Magnenat, and Laurent Keller. 2007. Evolutionary conditions for the emergence of communication in robots. *Current Biology* 17:514–519.

Hamilton, William Donald. 1964. The genetical evolution of social behaviour. *Journal of Theoretical Biology* 7:1–52.

Hamilton, William Donald. 1975. Innate social aptitudes of man: An approach from evolutionary genetics. In *Biosocial Anthropology*, ed. R. Fox, 133–153. London: Malaby Press.

Hauert, Sabine. 2010. Evolutionary synthesis of communication-based aerial swarms. PhD thesis, EPFL, Lausanne, Switzerland.

Hauert, Sabine, S. Leven, Jean-Christophe Zufferey, and Dario Floreamo. 2013. Beat-based synchronization and steering for groups of fixed-wing flying robots. *Proceedings of the International Symposium on Distributed Autonomous Robotics Systems*, 83:281–293.

Hauert, Sabine, Jean-Christophe Zufferey, and Dario Floreano. 2009a. Evolved swarming without positioning information: An application in aerial communication relay. *Autonomous Robots* 26 (1): 21–32.

Hauert, Sabine, Jean-Christophe Zufferey, and Dario Floreano. 2009b. Reverse engineering of artificially evolved controllers for swarms of robots. In *CEC'09 Proceedings of the IEEE Congress on Evolutionary Computation*, ed. A. Tyrrell, P. Haddow, J. Timmis, S. Smith, 55–61. Piscataway, NJ: IEEE Press.

Lehmann, Laurent, and Laurent Keller. 2006. The evolution of cooperation and altruism: A general framework and a classification of models. *Journal of Evolutionary Biology* 19:1365–1379.

Lehmann, Laurent, Laurent Keller, Stuart West, and Denis Roze. 2007. Group selection and kin selection: Two concepts but one process. *Proceedings of the National Academy of Sciences of the United States of America* 104:6736–6739.

Mitri, Sara, Dario Floreano, and Laurent Keller. 2009. The evolution of information suppression in communicating robots with conflicting interests. *Proceedings of the National Academy of Sciences of the United States of America* 106:15786–15790.

Mondada, Francesco, Giovanni C. Pettinaro, André Guignard, Ivo W. Kwee, Dario Floreano, Jean-Louis Deneubourg, Stefano Nolfi, Luca M. Gambardella, and Marco Dorigo. 2004. Swarm-bot: A new distributed robotic concept. *Autonomous Robots* 17:193–221.

Nolfi, Stefano, and Dario Floreano. 2000. *Evolutionary Robotics: The Biology, Intelligence, and Technology of Self-organizing Machines*. Cambridge, MA: MIT Press.

Sachs, Joel L., Ulrich G. Mueller, Thomas. P. Wilcox, and James J. Bull. 2004. The evolution of cooperation. *Quarterly Review of Biology* 79:135–160.

Sahin, Erol. 2005. Swarm robotics: From sources of inspiration to domains of application. In *Swarm Robotics (WS 2004)*, LNCS 3342, ed. E. Sahin and W. M. Spears, 10–20. Heidelberg: Springer.

Urzelai, Joseba, and Dario Floreano. 2001. Evolution of adaptive synapses: Robots with fast adaptive behavior in new environments. *Evolutionary Computation* 9 (4): 495–524.

Waibel, Markus, Laurent Keller, and Dario Floreano. 2009. Genetic team composition and level of selection in the evolution of cooperation. *IEEE Transactions on Evolutionary Computation* 13:648–660.

West, Stuart A., Stephen P. Diggle, Angus Buckling, Andy Gardner, and Ashleigh S. Griffin. 2007. The social lives of microbes. *Annual Review of Ecology Evolution and Systematics* 38:53–77.

West, Stuart A., Ashleigh S. Griffin, and Andy Gardner. 2007. Social semantics: Altruism, cooperation, mutualism, strong reciprocity and group selection. *Journal of Evolutionary Biology* 20:415–432.

10 Understanding Higher-Order Cognitive Brain Mechanisms by Conducting Evolutional Neuro-robotics Experiments

Jun Tani, Michail Maniadakis, and Rainer W. Paine

10.1 Introduction

In our everyday lives, we conduct complex cognitive behaviors without noticing their complexity. We can build up complex action programs for different goals without having to think about them, for instance, when going to a nearby convenience store or preparing a cup of coffee. An interesting point is that in many cases such goal-directed actions are compositional, meaning that entire actions can be decomposed into a set of reusable action units or behavior primitives. Arbib's (1981) motor schemata theory says that a diversity of actions can be generated by flexibly combining different behavior primitives stored in a memory pool. Some neuroscience researchers have considered that executive control in the prefrontal cortex (PFC) is responsible for this type of compositional cognitive operation (Fuster 1989). However, there is still much debate over the details of neural mechanisms and architectures that might underlie this executive control. In this chapter we propose that evolutionary robotics (ER) methods can be useful tools in helping to illuminate this question by allowing us to explore it without having to make too many assumptions, as explained in earlier chapters of this book (particularly chapter 2).

It is widely considered that executive control in the PFC is not just for combining or sequencing behavior primitives in goal-directed actions, but is also involved with other higher-order cognitive tasks such as monitoring, evaluating, inhibiting, and sustaining other ongoing processes in a role that is analogous to that of a computer operating system. A prototype test for examining such executive control capabilities in humans is the Wisconsin Card Sorting Test (WCST) (Berg 1948; Milner 1963), for which subjects are invited to discover and apply a given card-sorting rule based on reward and punishment feedback. During the experiment, the rule is changed unpredictably by the experimenter and must be rediscovered. Although it is said that this test requires working memory, for the rule in effect at any given moment, in the dorsolateral prefrontal cortex (DLPFC) (Mansouri, Matsumoto, and Tanaka 2006); and conflict monitoring in the anterior cingulate cortex (ACC) along with reward/punishment feedback (Kerns et al. 2004), the related local functionality in the PFC is still a matter of ongoing debate (Stoet and Snyder 2009).

Existing computational modeling studies on the WCST tend to impose discrete and algorithmic computational processes on the models based on the common assumption that, although posterior cortices can be characterized as fundamentally analog systems, the PFC has a more discrete, digital character (O'Reilly 2006; Dayan 2007). Some studies (Dehaene and Changeux 1991; Stemme, Deco, and Busch 2007) have employed a local and discrete representation in neural network models where currently adopted rules are represented by the activation of the corresponding local units. Rougier and O'Reilly (2002) have proposed an on-off type of gating operation acting on working memory for storing information about currently adopted rules.

We see similar ideas of local representation and their external manipulations in constructing neuronal models for compositional action generations. Tani and Nolfi (1999) once proposed a hierarchical model in which each behavior primitive is stored in a corresponding local modular network at the lower level whereas a higher level network sequentially selects activated local module one by one by opening and closing gates associated with the local modules. Haruno, Wolpert, and Kawato (2003) also considered a similar model. Although an idea of locally representing rules or primitives as manipulable objects for the higher executive control level is easily understandable from the computational view, it is not yet clear that real biological brains actually perform in this manner. In particular, electrophysiological experiments on monkeys trained to perform WCST analogs showed that assemblies of DLPFC cells encode rules through different distributions of dynamically changing firing activity (Mansouri, Matsumoto, and Tanaka 2006).

Rather than hand coding specific computational mechanisms in model networks, our proposal is to look at what sorts of neural mechanisms could appear by means of self-organization of internal structures in simple neural network models through their adaptation to achieve target tasks by utilizing the evolutionary robotics scheme (Koza 1992; Cliff, Harvey, and Husbands 1993; Nolfi and Floreano 2000). More specifically, general types of neural network models with recurrent connectivity are evolved to perform robotics tasks involving two classes of higher-order executive control functions using a standard genetic algorithm to search for optimal synaptic weights maximizing the task fitness. One task involves a compositional goal-directed action generation and the other is concerned with a rule-switching behavior similar to WCST. If the same neural mechanism consistently appears for each robotics task in repeated evolutionary runs, comparable principles might be applicable also in real brains. The following sections will briefly describe the simulation experiment for each target task one by one. For further details of each experiment refer to (Paine and Tani 2005; Maniadakis and Tani 2009).

10.2 Goal-Directed Compositional Action Generation

An exploratory navigation task of a simulated mobile robot was considered for investigation of possible neural mechanisms for goal-directed compositional action generation.

10.2.1 Model

A simulated mobile robot equipped with eight proximity sensors and two motor-driven wheels explores a maze environment shown in figure 10.1. The task of the robot is to find navigation paths reaching as many different goals as possible from a start position. This navigation task can be deconstructed into two levels of system functions. The first level should deal with collision-free maneuvering, going straight along a corridor, and turning left or right at corners. The second level should deal with sequencing the turning at corners in order to reach a set of different goals. The goal of the study is to understand how two such levels of functions can be self-organized in neural networks from scratch without showing explicit cues. Our navigation task is unique compared to other navigation tasks conducted by other groups (Ziemke and Thieme 2002; Nolfi 2002; Blynel and Floreano 2003) because our task requires the robot to deal with multiple goals. It is expected that this requirement will force the neural system to organize to make use of compositionality.

The robot is implemented with a fully connected CTRNN (continuous-time recurrent neural network) that is evolved by a genetic algorithm (GA). The activation dynamics of each neuronal unit is given by

$$\tau \dot{u}_i = -u_i + \sum w_{ij} a_j \tag{10.1}$$

$$a_i = \sigma(u_i + b_i) \tag{10.2}$$

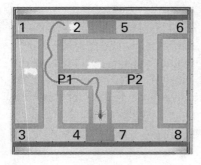

Figure 10.1
A simulated mobile robot learns ways to reach eight different goals starting from the home position.

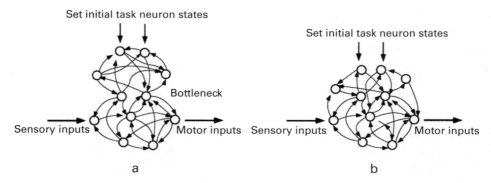

Figure 10.2
(a) CTRNN with a bottleneck and (b) standard CTRNN.

Where u is the activation, a the output, w a connection weight, and b a bias, τ a time constant and $\sigma(x) = 1/(1+e^{-x})$ is a standard sigmoid function. We tested two types of CTRNNs as shown in figure 10.2. Figure 10.2a is called a "bottleneck" network since the information flow between the top and bottom levels is narrowed into a bottleneck. (The neural activations can propagate to the other level only through the bottleneck neurons [BN].) The bottleneck CTRNN has five neurons in the lower part, two BNs, and four neurons in the upper part. There are two so-called task neurons (TN) in the upper part whose functions will be explained later. All neurons in the lower part receive eight proximity sensor inputs and output to two motor neurons, driving left and right wheels, through synaptic connections. Figure 10.2b is a standard CTRNN consisting of nine neurons including two TNs. All neurons receive eight sensory inputs and output to two motor neurons.

We employed the ideas of initial sensitivity to generate combinatorial action sequences in the current task. The idea in the current setting is that the robot can reach different goals depending on the initial state values set in the TNs shown in figure 10.2. In the evolutionary process, a set of the initial state values in the task goal neurons evolves, along with the synaptic weights and the biases. The time constant τ for each neuronal unit is also evolved. The fitness function is designed to increase the number of different goals reached with the set of evolved initial state values. We repeated the evolutionary runs twenty times for both types of networks for statistical comparisons of their performances. Each evolutionary run is conducted for 200 generations with an eighty-robot population per generation.

10.2.2   Results
Our results showed that the best performance is obtained in the bottleneck network. In twenty evolutionary runs, the average number of different goals reached was 5.1

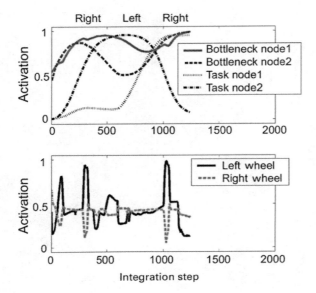

Figure 10.3
Neuronal activity for a right-left-right turn sequence in the bottleneck network. Top: neuronal activity of bottleneck and task neurons, respectively; bottom: activities of motor output nodes.

for the bottleneck CTRNN, and 2.3 for the standard CTRNN. The bottleneck CTRNN found five or more goals on fourteen of twenty runs. The standard CTRNN found them on only six runs.

The temporal neuronal activation profiles for an evolved bottleneck network, which found six different goals, are shown in figure 10.3. The profiles correspond to a right-left-right turn sequence, starting from the home position that reaches goal 6 of figure 10.1.

The top row shows the activation profiles of two TNs and two BNs (see figure 10.2a). The bottom row shows the profiles of the two motor output neurons in the lower part of the network. Observe that the motor outputs show much faster dynamics than those of the TNs and BNs. Actually, we found that the time constants for the motor neurons evolve to be much faster than those of the TNs and BNs in all successful evolutionary runs. The activation profiles of the BNs correlate with right and left turns, denoted by labels in the top figure. For the right turn, both BNs have high activation values, while BN-2 takes a low value and BN-1 slightly decreases for the left turn. TN-2 shows a similar type of encoding to the BNs, while the dynamic profile of TN-1 seems uncorrelated with the turn sequence. These profiles suggest that certain structures in the levels are self-organized in the bottleneck network. The following analysis examines such structures.

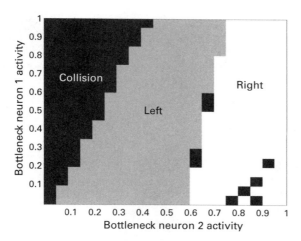

Figure 10.4
Phase space analysis for two bottleneck neurons.

First, functions of the BNs were examined. We constructed a phase space analysis for the BNs, focusing on the cornering behavior at the T branch. Figure 10.4 indicates how the cornering behavior varies when the activation of two BNs are clamped externally to various values. It is observed that the BNs' activation space is divided into three regions, gray, white, and black, which correspond to left turns, right turns, and collisions with the walls, respectively. It is considered that the BNs' activation states encode the behavior primitives of turning left or right in branches.

Next, we constructed a phase space analysis for the task neurons, initial states, focusing on their possible encoding for the turning sequences. The results can be seen in figure 10.5, where the regions in the initial state space that reach different goals are labeled by the corresponding turn sequence, for example, LRR for a left-right-right turn sequence. The turn sequence is denoted by number in the plot (see the legend on the right in the figure).

It is observed that the sequence patterns are arranged in clusters in the TN initial state space. First, the space is grossly clustered based on the first turn direction, left or right, of the movement sequence, as shown by a thick solid line in figure 10.5. Each of these two clusters is then further divided into topologically ordered subclusters, depending on the second turn direction of the movement sequence, as shown by a solid line. These subclusters are still further divided into smaller clusters, depending on the third turn as shown by the dashed lines. These smallest clusters neighbor each other and share the first two turns of their sequences in common. In other words, the turn sequences are hierarchically ordered into progressively smaller regions of the

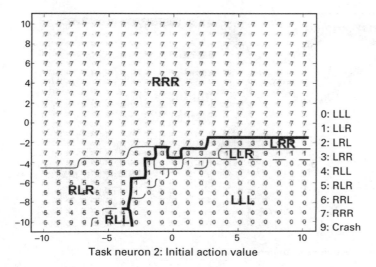

Figure 10.5
Phase space analysis for task neuron initial states. Plotted numbers correspond to turn sequences as in the legend on the right.

initial TN activity space as additional turns are added. As the complexity of the movement sequence increases, so too does the initial sensitivity to the TN activities.

In order to clarify the functional roles of the upper level of the bottleneck network, we observed the activities of the upper-level neurons while they were decoupled from the lower-level ones—that is, disconnecting all the synaptic connections from the lower-level neurons to the BNs. It turned out that the activities over time of the TNs and BNs are mostly the same as the original ones provided that the same initial states are set in TNs. Compare the disconnected case shown in figure 10.6 with the original one shown in figure 10.3 for reaching goal 6.

The results imply that the whole network was evolved such that the upper level generates top-down internal images or plans for achieving the goals without accessing the sensory inputs, and that the lower level deals with actual maneuvering control of the robot based on the plans. More specifically, the upper level generates the top-down anticipation of how the BNs' states should develop based on the goal information encoded in the initial states of the TNs while the states of the BNs activate the behavior primitives of turning left or right in sequences in the lower level.

Finally, we consider why the case of the standard fully connected CTRNN cannot evolve successfully as compared to the bottleneck case. It is assumed that evolving different dynamic functions with different time constants is difficult within a single

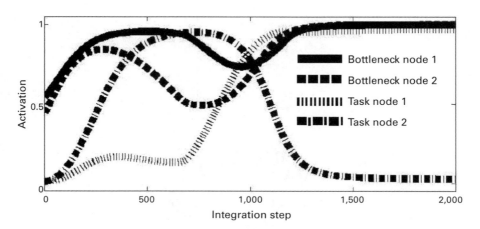

Figure 10.6
The profiles of BNs and TNs activities in upper-level disconnected case (reaching to goal 6).

fully connected network because it would cause too much interference among them. In the bottleneck case, fast and slow dynamics can be evolved more easily by having less interference with each other since they are segregated by the bottleneck of the network. Our experiments showed that a class of level-structured functions can be evolved provided that adequate topological constraints such as bottlenecks or hub-like connectivities are imposed on the network. This should be also true for real brains that are known to have partial connectivity segregation between the PFC and the posterior cortices. Our speculation is that such partial segregation would make the PFC a special place suitable for the executive control of other brain regions.

10.3 Rule-switching Task

We considered a robotic rule-switching task similar to the WCST to examine possible neural mechanisms for executive control of rule switching.

10.3.1 Model

A mobile robot equipped with range sensors for obstacles and light sensors navigates a T-maze environment (see figure 10.7) by following the current rule set by an experimenter. Two light sources are located on the left and right sides of the walls near the start position at the bottom of the T-maze. In each trial, the robot (starting from the start position) perceives light from either side and proceeds to determine the branching (left or right) depending on the currently adopted rules and the side from which the light was perceived. If the robot reaches either side of the T-wings within 165 simulation steps by following the current rule, it receives no punishment signals and the trial

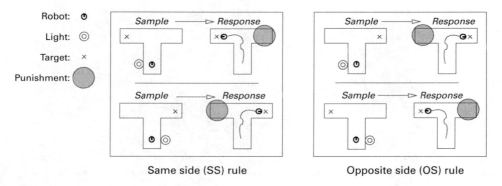

Robot: ☉
Light: ◎
Target: ×
Punishment: ●

Same side (SS) rule   Opposite side (OS) rule

Figure 10.7
A graphical interpretation of the two behavioral rules used in our experiments. Light samples are depicted with double circles, each target location is depicted with an X, and the punishment area is depicted with a gray circle.

is regarded as successful. However, it receives a punishment signal if it fails. After each trial, the robot is moved back to the start position by the experimenter. It should be noted that the internal neural dynamics are continued without being reset when the position is reset. In the current experiment, two rules are considered, namely the same side (SS) rule and the opposite side (OS) rule. In SS, the trajectory of the robot branches in the same direction as that of the light source, whereas in OS it branches in the opposite direction, as shown in figure 10.7.

The robot acquires each of the preceding two rules and furthermore learns that the currently adopted rule should be switched to the other rule if a punishment signal has been received. It should be noted that there is no explicit cue for the switching of rules and it is unpredictable when the switching will take place. After repeating the same rule for thirteen trials, there is a chance that the experimenter will switch the rule.

Two types of CTRNN architectures with/without the bottleneck connectivity were evolved to achieve the task and their performances were compared (figure 10.8). Both networks, consisting of the same number of neural units ($N = 15$), have connections to the same input in the form of range sensors, light sensors, and punishment signals and the same output in the form of two wheels driven by a motor. The connection weights in these two networks were evolved by using a standard genetic algorithm (GA) with a fitness function based on the success rate. In this model, the time constant τ is set to a constant value for all neural units.

10.3.2   Results
We examined the robot performances for both the fully connected and the bottleneck CTRNN, conducting ten independent evolutionary runs for each network type. For

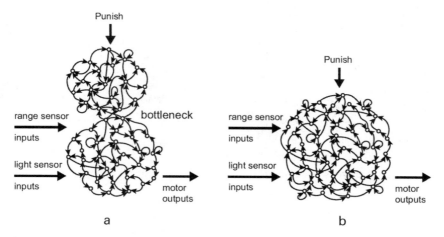

Figure 10.8
(a) CTRNN with bottleneck and (b) fully connected CTRNN employed for the rule-switching task.

the case of the bottleneck CTRNN, eight out of the ten evolutionary processes converged successfully producing controllers capable of accomplishing the given tasks. However, only three out of the ten evolutionary processes converged successfully for the case of the fully connected CTRNN. These results are analogous to the one in the previous experiment. Due to the significantly better performance of the bottleneck CTRNN, for the rest of the section we will concentrate our study on the results from the bottleneck CTRNN case.

The behavior of the robotic agent for one representative bottleneck CTRNN is demonstrated in figure 10.9. During trials 1–4 the robot is successfully following the opposite side (OS) rule. Then, in the fifth trial the rule is unexpectedly changed to same side (SS) rule, and the agent produces a wrong response driving in the punishment area. At that time, the agent understands that its current response strategy is not correct anymore, and it adopts another response rule. Accordingly, it adopts the SS rule, responding successfully for the next eleven trials, avoiding punishment signals. The rule is unexpectedly changed again in trial 17, where the robot gives a wrong response driving again into the punishment area. This time it takes two trials for the agent to revert back to the OS rule. After that, the agent gives correct responses in the subsequent trials.

Interestingly, it was found that robot paths are significantly correlated with the currently adopted rule. For example, every time the robot turns left according to the SS rule it follows very similar trajectories (compare trials 8, 9, 15 in figure 10.9). The same is also true when it turns to the right for the same rule (see trials 12, 13, 16 in figure 10.9). A similar relationship can be observed for the paths of the OS rule

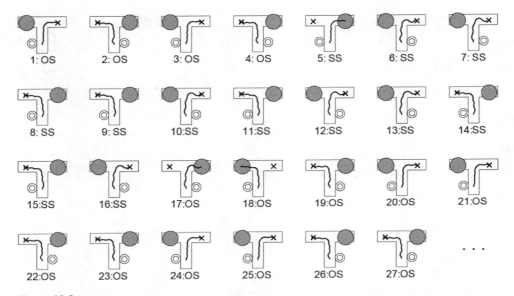

Figure 10.9
The behavior of the agent in a sequence of trials. The light is depicted with a double circle, the goal position is depicted with an X, the punishment area is depicted with a gray circle, while the robot path is depicted with a black line starting from the bottom of the T-maze.

(compare for example right turns in trials 3, 20, 21, and additionally compare left turnings in trials 19, 26, 27). However, by comparing same side turnings under different rules, we can see different trajectory characteristics (for example comparing trials 12, 13 with trials 24, 25). This means that robot trajectories are somehow involved in distinguishing the two rules. In other words, the CTRNN controller takes advantage of its embodiment and environmental interaction in generating specific maneuvering under the currently adopted rule.

We turn back now to the results shown in figure 10.9, and their relationship to the neural activities shown in figure 10.10. We previously commented that we observed very similar behaviors every time the robot responds to the same side, following a given rule (e.g., for all left turns of the SS rule). Additionally, very similar activation patterns are observed in the higher- and lower-level neurons in each one of these cases. This means that the composite CTRNN controller has stored internally a set of different behavioral procedures, which are properly selected and expressed, based on the activity of the higher-level neurons and the sensory light input. This emergent function is similar to the one in parametric bias neurons (Tani and Ito 2003; Nishimoto and Tani 2004), which has been shown to facilitate storing and recalling a set of behaviors to the same network.

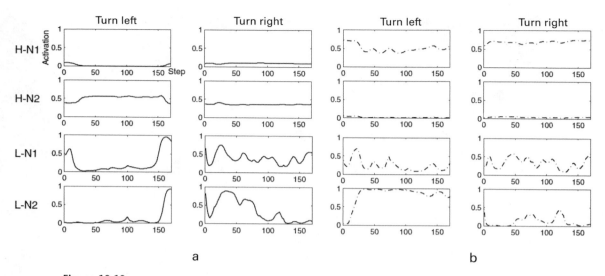

Figure 10.10

The activation of two higher-level (H-N1, H-N2) and two lower-level (L-N1, L-N2) neurons when the agent follows (a) the SS rule and (b) the OS rule. SS is depicted with a solid line while OS is depicted with a dashed line.

After conducting attractor analysis, neural characteristics correlated to SS and OS rules were identified in both the higher and the lower part of the CTRNN. Specifically, for each rule, we asked the agent to perform 1,000 trials with a randomly located light source either on the left or right side at each trial. We observed that after an initial transient period, the agent's behavior always converges to the correct response strategy by utilizing the punishment feedback for each rule, implying that rule-based attractors have emerged in the network dynamics. To confirm this, the phase plots for the higher- and lower-level neurons for each rule are shown in figure 10.11. For each rule, the same shape of attractor appears in the plot in repeated examinations. As was expected, for each rule a distinct invariant set of dynamically changing trajectories is observed in the higher-level neural activity. Additionally, we can see that distinct invariant sets have also emerged in the lower level. It can be seen that the shapes of the invariant sets in the higher level are much more compact than the ones in the lower level. This implies that the higher level functions as a working memory to memorize the currently adopted rules with abstraction, whereas the lower level takes care of the details of sensorimotor control by following the adopted rules. Then, the rule switch is enabled by a transition of dynamic state from one attractor to the other as triggered by the punishment feedback. Because the same

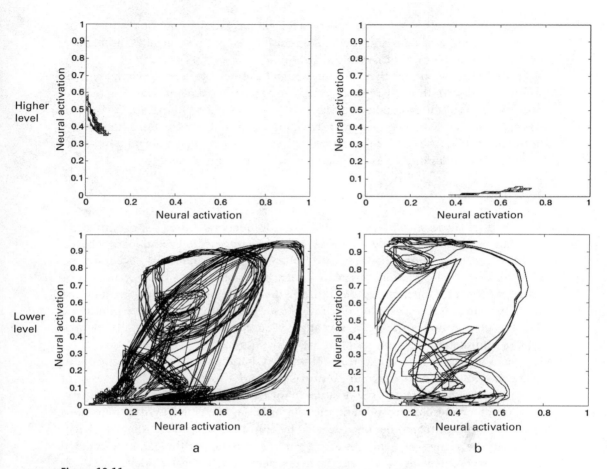

a   b

Figure 10.11

The phase plots of higher- and lower-level neural activity when the agent follows (a) the SS rule and (b) the OS rule. In the figures of the first row, the axes *x* and *y* correspond to the activity of neurons H-N1 and H-N2, while in the figures of the second row the axes *x* and *y* correspond to the activity of neurons L-N1 and L-N2.

dynamical mechanism has been observed in all successful evolutionary runs, the attractor encoding of rules and their state transition might be a general mechanism for the executive control of the rule switching.

Furthermore, we found that the same model can achieve a more complex rule-switching task with three rules by self-organizing the same attractor switching mechanism as the result of evolution (see Maniadakis and Tani 2009). The only difference was that the rule switching takes a greater number of trials for the transition period in the three rules case. This is natural because there is a potential ambiguity in selecting alternative rules in the case of punishment for the current rule.

10.4   Discussion

This chapter has described how executive control functions can be self-organized for two different higher-order cognitive tasks by conducting neuro-evolutionary robotics experiments. In the first experiment, two different levels of functions have self-organized as a result of evolution: one is to sequence behavior primitives of either left or right branching for different goals by utilizing the initial sensitivity characteristics; the other is to realize the sensorimotor control associated with the behavior primitives. The executive control in this task was attained successfully by utilizing a bottleneck type of information flow constraint in the model network.

The results of the second experiment indicated that the same bottleneck constraint also enhances the performance in executive control for rule switching. As a result of evolution, multiple attractors self-organize in the network dynamics where each attractor embeds a corresponding rule and the rule switching is enabled by state transitions from one attractor to another one triggered by punishment feedback. The neural activation in the higher level encodes abstract information about the currently adopted rules as a working memory, whereas the neural activation in the lower level takes care of sensorimotor-level control for the rules. It is noted that Ziemke (1996) showed that an RNN controller evolved by a genetic algorithm can achieve some context switching tasks that are analogous to our rule-switching task. However, because Ziemke's task does not involve the complexity of rules but is simply involved with sensory situations, his task does not require the organization of any functional hierarchy like that shown in our results.

One interesting finding was that the neural internal representation for the behavior primitives or the rules achieved by evolution turn out to always be distributed rather than local. The distributed representation is considered to have some advantages over the local one. First, the distributed representation can be more compact than the local one because if there are common structures between different primitives or rules, they can be shared. In other words, the distributed representation can represent what is distinct or common among others efficiently in one body of memory structure and therefore it can achieve generalization more easily than the local one. This cannot be afforded by the local representation scheme in which all memory items are isolated in local modular networks. Second, there is a stability problem in the local representation scheme (Tani et al. 2008). The instability in the local representation scheme originates from the fact that the currently activated primitives or rules have to match exactly with one of a finite number of stored memory modules. If a near miss takes place in the matching, the winner-take-all dynamics of selecting the most appropriate memory module tends to be unstable. In the case of the distributed representation, the primitives or rules in the lower level in the bottleneck architecture are represented with sort of "elasticity" along with smooth changes of the bottleneck neuron-activation values. For example, exact trajectories for left turning or right turning in the first experiment can modulate with analog patterns of bottleneck neuron activations. When the way of turning left or right has to be slightly modulated at each instance, this modulation would cause only minor modulations in the activations of the bottleneck neurons in a bottom-up manner. It is strongly suspected that the evolutionary processes tend to select a distributed representation rather than a local one because of these advantages.

It is, however, interesting to see that the evolutionary processes are likely to generate distinct locality in representing functional levels in both experiments. This might be because there are no shared structures between the memory contents at different levels, that is, behavior primitives and their sequencing or working memory of current rules and their sensorimotor details. If the network attempts to represent contents belonging to different levels in a distributed manner, its performances become worse as has been shown in the results of the fully connected network cases.

Many computer scientists have considered that higher-order cognition dealing with some hierarchy should involve combinatorial symbolic computation processes. This is, however, not always true because dynamical systems defined in continuous time and space can also exhibit combinatorial complexity by utilizing their nonlinear systems characteristics such as parameter bifurcation and sensitivity to initial conditions relating to chaos (Wiggins 1990; Crutchfield 1989; Tani and Fukumura 1995). Actually our evolutional neuro-robotics experiments have suggested the possibility that even higher-order cognition involved with the executive control of primitives and rules can be realized by adopting analog neural dynamics. In this situation, the top-down executive control function assumed to exist in the prefrontal cortex can have a more natural interaction with the posterior cortex, which is responsible for the bottom-up sensorimotor processes, because both sides share the same metric space of analog dynamical systems, as our group has discussed for more than a decade (Tani 1996, 1998; Yamashita and Tani 2008). In fact, our simulation results in the rule switching task have shown that the sensorimotor level embodiment positively participated in the encoding of higher-level rules.

Evolutionary neuro-robotics could provide a new approach to computational neural modeling studies. On the one hand, conventional neural modeling studies usually begin with computational frameworks predefined in detail by the researchers themselves. Then, the purpose of the simulation experiments is to validate the performances of the models by identifying possible parameter ranges. On the other hand, the evolutional neuro-robotics studies start with computational experiments on dynamic interactions among neural systems, bodies, and environments to achieve specific cognitive tasks without making assumptions about the exact mechanisms. Such computational experiments could show us novel neural mechanisms to solve given cognitive tasks that the experimenters could not have imagined beforehand. If such neural mechanisms appear repeatedly as the results of evolving simple neural network models, the same principle may well be applicable also in real brains.

We, however, admit that the current CTRNN model is too simple to support the neuroscientific reality even at the minimum level. Future research will consider the inclusion in the models of known neurophysiological constraints such as anatomical connectivity and neurochemical substrates.

References

Arbib, M. 1981. Perceptual structures and distributed motor control. In *Handbook of Physiology: The Nervous System, II. Motor Control*, ed. V. B. Brooks, 1448–1480. Cambridge, MA: MIT Press.

Berg, E. 1948. A simple object test for measuring flexibility in thinking. *Journal of General Psychology* 39:15–22.

Blynel, J. 2003. Evolving reinforcement learning-like abilities for robots. In *Proceedings of the 5th International Conference on Evolvable Systems: from Biology to Hardware*, ed. A. Tyrell, P. C. Haddow, and J. Torresen, LNCS 2606, 320–331. Berlin: Springer.

Cliff, D., I. Harvey, and P. Husbands. 1993. Exploration in evolutionary robotics. *Adaptive Behavior* 2 (1): 73–110.

Crutchfield, J. 1989. Inferring statistical complexity. *Physical Review Letters* 63:105–108.

Dayan, P. 2007. Bilinearity, rules, and prefrontal cortex. *Frontiers in Computational Neuroscience* 1:1–14.

Dehaene, S., and J. Changeux. 1991. The Wisconsin card sorting test: Theoretical analysis and modeling in a neuronal network. *Cerebral Cortex* 1:62–79.

Fuster, J. 1989. *The Prefrontal Cortex*. New York: Raven Press.

Haruno, M., D. Wolpert, and M. Kawato. 2003. Hierarchical mosaic for movement generation. *International Congress Series* 1250:575–590.

Kerns, J. G., J. D. Cohen, A. W. MacDonald III, R. Y. Cho, V. A. Stenger and C. S. Carter. 2004. Anterior cingulate conflict monitoring and adjustments in control. *Science* 303:1023–1026.

Koza, J. 1992. Evolution of subsumption using genetic programming. In *Proceedings of the First European Conference on Artificial Life,* ed. F. J. Varela and P. Bourgine, 110–119. Cambridge, MA: MIT Press.

Maniadakis, M., and J. Tani. 2009. Acquiring rules for rules: Neuro-dynamical systems account for meta-cognition. *Adaptive Behavior* 17 (1): 58–80.

Mansouri, F., K. Matsumoto, and K. Tanaka. 2006. Prefrontal cell activities related to monkeys' success and failure in adapting to rule changes in a Wisconsin card sorting test analog. *Journal of Neuroscience* 26 (10): 2745–2756.

Milner, B. 1963. Effects of different brain lesion on card sorting. *Archives of Neurology* 9:90–100.

Nishimoto, R., and J. Tani. 2004. Learning to generate combinatorial action sequences utilizing the initial sensitivity of deterministic dynamical systems. *Neural Networks* 17:925–933.

Nolfi, S. 2002. Evolving robots able to self-localize in the environment: The importance of viewing cognition as the result of processes occurring at different timescales. *Connection Science* 14 (3): 231–244.

Nolfi, S., and D. Floreano. 2000. *Evolutionary Robotics: The Biology, Intelligence, and Technology of Self-organizing Machines.* Cambridge, MA: MIT Press/Bradford Books.

O'Reilly, R. 2006. Biologically based computational models of high-level cognition. *Science* 314:91–94.

Paine, R., and J. Tani. 2005. How hierarchical control self-organizes in artificial adaptive systems. *Adaptive Behavior* 13 (3): 211–225.

Rougier, N., and R. O'Reilly. 2002. Learning representations in a gated prefrontal cortex model of dynamic task switching. *Cognitive Science* 26:503–520.

Stemme, A., G. Deco, and A. Busch. 2007. The neuronal dynamics underlying cognitive exibility in set shifting tasks. *Journal of Computational Neuroscience* 23:313–331.

Stoet, G., and L. Snyder. 2009. Neural correlates of executive control functions in the monkey. *Trends in Cognitive Sciences* 13 (5): 228–234.

Tani, J. 1996. Model-based learning for mobile robot navigation from the dynamical systems perspective. *IEEE Transactions on Systems, Man, and Cybernetics (B)* 26 (3): 421–436.

Tani, J. 1998. An interpretation of the "self" from the dynamical systems perspective: A constructivist approach. *Journal of Consciousness Studies* 5 (5–6): 516–542.

Tani, J., and N. Fukumura. 1995. Embedding a grammatical description in deterministic chaos: An experiment in recurrent neural learning. *Biological Cybernetics* 72:365–370.

Tani, J., and M. Ito. 2003. Self-organization of behavioural primitives as multiple attractor dynamics: A robot experiment. *IEEE Transactions on Systems, Man, and Cybernetics, Part A* 33 (4): 481–488.

Tani, J., R. Nishimoto, J. Namikawa, and M. Ito. 2008. Codevelopmental learning between human and humanoid robot using a dynamic neural network model. *IEEE Transactions on Systems, Man, and Cybernetics* 38 (1): 43–59.

Tani, J., and S. Nolfi. 1999. Learning to perceive the world as articulated: An approach for hierarchical learning in sensory-motor systems. *Neural Networks* 12:1131–1141.

Wiggins, S. 1990. *Introduction to Applied Nonlinear Dynamical Systems and Chaos*. New York: Springer-Verlag.

Yamashita, Y., and J. Tani. 2008. Emergence of functional hierarchy in a multiple timescale neural network model: A humanoid robot experiment. *PLoS Computational Biology* 4:e1000220.

Ziemke, T. 1996. Towards adaptive behaviour system integration using connectionist infinite state automata. In *From Animals to Animats 4*, ed. P. Maes, 145–154. Cambridge, MA: MIT Press.

Ziemke, T., and M. Thieme. 2002. Neuromodulation of reactive sensorimotor mappings as short-term memory mechanism in delayed response tasks. *Adaptive Behavior* 10 (3/4): 185–199.

11 Incremental Evolution of an Omni-directional Biped for Rugged Terrain

Eric D. Vaughan, Ezequiel A. Di Paolo, and Inman Harvey

11.1 Introduction

Natural evolution has produced humans that can walk and talk, without any explicit design process; Darwinian evolution has taken the role of the Blind Watchmaker. The process took some four billion years overall, with inconceivably immense resources and plenty of dead ends. It was not aimed, we may assume, at the end goal of walking, talking humans; there are certainly plenty of viable species that neither walk nor talk.

A human designer, aiming to replace at least part of the explicit design process by an evolutionary robotics (ER) methodology based on Darwinian evolution, will have comparatively tiny resources and a limited timeframe. There are many possible motives for using ER (Harvey et al. 2005), and the one presented here can be called an engineering motivation: to design a mechanism for a specific application. The use of ER will only be justified to the extent that it produces better results than can be expected through conventional design methods. The human engineer will have a focused goal, and will want to apply every trick that can be found to speed up the evolutionary process. There will typically be a continuing interplay between the role of "blind" ER and the vision of the engineer as a way of iteratively finding the evolutionary pathways toward the desired solution (chapter 4, this volume); we present here one case study of how this can work.

The engineering goal in our case study is the design of efficient and robust machines for bipedal walking in any direction on both flat and irregular surfaces. Bipedal robots have the potential to replace or assist humans in the types of terrain that they use, including rugged surfaces outdoors and steps and stairs indoors. It turns out that replicating human walking is a challenge. One reason human locomotion is so efficient is that it leverages passive dynamics to reduce energy consumption and uses the elastic nature of tendons to store and release energy; these considerations have been missing from traditional robot design. Here we present one part of a larger body of work undertaken by the first author of this chapter in doctoral research on the development of bipedal walking (Vaughan 2007). The results presented include the successful coordination and control of many more degrees of physical freedom (up to thirty-five) than are typically tackled by conventional design methods; this endorses the effectiveness of the ER methodology. Going beyond the domain of bipedal walking, we consider that the ongoing interplay between engineering vision and ER methods in incremental design, as illustrated in this case study, may have valuable lessons for a wider audience.

11.2 Empowering ER through Incremental Design

Natural evolution has clearly been incremental, at both micro- and macroscales, with current generations altering and extending the design achievements of earlier ones. Human design methods are likewise often incremental. The Wright brothers started by adding elements of active control to kites, so as to produce unpowered dynamic flying machines—gliders. Only then did they go on to adding power and increasing the sophistication of the controls. Our current aircraft can trace their ancestry through continued incremental improvements from those early days.

This work in this chapter takes inspiration from the Wright brothers, but applied to walking. McGeer (1990) showed that a simple set of jointed legs, in proportions similar to human legs, with knees, could walk down a slope with no power other than that provided by gravity, and no control other than that provided by the pendulum dynamics. This passive dynamic walker (PDW) can play the part of the Wrights' glider, to which power and increasingly sophisticated levels of control are to be added. This implies an incremental design pathway that can be well suited to ER. Ways in which an evolutionary algorithm can be applied to such an incremental process, and the use of incremental methods in ER, were proposed in Harvey 1992 and Harvey, Husbands, and Cliff 1993, and a body of work following from this. Prior work applying incremental evolution specifically to walking robots includes Kodjabachian and Meyer 1998.

Brooks's (1991) "subsumption architecture" also advocates an incremental design approach. Though it is explicitly inspired by the incremental aspects of natural evolution, design by hand is used throughout. The emphasis is on building complete robots that initially have simple behaviors, and then adding extra functionality to enable extra layers of behavior and more sophistication. At each successive stage, the robot has to function successfully in the real world at its currently expected level, and only after this is achieved will the next stage be added. We can consider several potential advantages to this incremental approach, over and beyond the fact that it follows good engineering principles of iterative development and testing.

First, it breaks down what may be one very large design problem into many smaller ones, each of them individually more tractable. Since limited resources may well make cracking the big problem in one go unlikely or impossible within the available timescale, the achievement of some intermediate stepping-stones can be a better result. Second, intuitions supported by anecdotal evidence (and meriting more principled investigation) suggest that it may well often consume fewer resources to achieve an ambitious end goal via intermediate stepping-stones than it would to attempt it in one go, even if the engineer learned nothing new during the process. Third, as will be indicated in the examples that follow, it is likely that the engineer will indeed learn of significant new factors during the intermediate stages, and this can lead to recognition of stumbling blocks, improvements in choosing what the next stepping-stone might be, and the possibility of making available new and appropriate resources for the next stage of design. The ER example here illustrates an ongoing collaboration between the engineer and the evolutionary process; often the former is providing the broad brush strokes outlining the direction the next design stage should take, while the latter is providing the essential detail by juggling the parameters of a highly complex system so as to coordinate the different parts. Here the coordination needs to be between neural and physical dynamics. Walking involves real-time dynamics, so it is natural that the control system being involved should also cope with real-time dynamics. This consideration influenced the choice of continuous-time recurrent neural networks (CTRNNs) (Beer 1995), to be described in more detail.

Most forms of evolutionary algorithm will handle incremental evolution satisfactorily, so the details of the GA (genetic algorithm) used here are not significant. One observation to note is that, at each successive stage of evolution, the population will be based on that which succeeded at a preceding stage, and hence will tend to be always quite genetically converged rather than initially randomized. This has some implications for the evolutionary dynamics (Harvey 2001).

11.3 Staged Evolutionary Design of Walkers

In previous work (Vaughan et al. 2004a; Vaughan 2007), passive dynamics were explored in physical simulations using staged evolutionary design (SED)—in this case, three stages. In the first stage a ten-degrees-of freedom-machine was created with hips and ankles; it could walk down a gentle slope unpowered by optimizing the physical properties of the body. In the second stage a simple neural network was hand-designed and coupled with a central pattern generator. In the final stage sensor input was added to the network and the slope was lowered to a flat surface over many generations. This machine showed that efficient walking attractors can be generated in the body itself and it is possible to vary their range from sloped surfaces to flat ones by adding simple stabilizing neural networks. In following work (Vaughan, Di Paolo, and Harvey 2004b, Vaughan 2007), some of the weaknesses of this model were addressed, specifically its lack of a weighted torso and inability to walk backward. Using a more complex network and prior knowledge the sloped platform stage was bypassed and the machine learned to walk directly on a flat surface.

In this chapter, a more sophisticated 3D bipedal machine is developed that can walk unsupported in any direction on both flat and irregular surfaces. First we discuss and explore simple models and some basic principles of walking and balance. This is followed by a description of the body and networks used. We develop the machine in three stages demonstrating the power of the incremental methodology.

In stage one, we develop a simple planar machine based on previous work that can walk forward and backward on a flat surface. Manoonpong et al. (2007) have explored the addition of neural circuits to a passive dynamic walker in the planar case by using synaptic plasticity to achieve adaptive control, but with significant differences from the approach described here. Through observation, we add improvements to increase the machine's performance and to allow it to walk on rugged terrain. In stage two, we add a lateral control system allowing the machine to walk unrestrained in three dimensions. We test the machine on flat and rugged surfaces, where it shows the ability to walk at different speeds and make dynamic quick movements in response to the environment. In stage three, we examine and discuss some implications of walking with ankles and flat feet. At the end of the chapter, we discuss preliminary experiments with a spine, passive arms, and extra hip joints. These are difficult problems that have not been studied previously but can be approached using the SED methodology.

11.4 Walking Revisited

Following the incremental strategy, we reexamine the same problem of robust walking but at each stage taking experience from earlier work (Vaughan, Di Paolo, and Harvey 2004a; Vaughan, Di Paolo, and Harvey 2004b, Vaughan 2007). Previous models are scaled up to a machine with thirty-five degrees of freedom with a flexible spine. At this point it is beneficial to revisit what has been learned from previous work. In particular, we return to the process of walking to come up with a strategy for scaling up to a machine that could challenge trajectory-based machines such as Honda's Asimo. We focus in particular on three concepts in walking: *foot placement*, *foot passing*, and *weight balancing*.

Foot placement Walking can be thought of as controlled falling, whereby the legs consistently break the fall of body mass on each step. Generally this is controlled by foot placement. The larger the angle between the hip and the leg when the foot strikes the ground, the greater the decrease in the body's velocity. Smaller angles can act to increase velocity by failing to reduce the machine's fall (Raibert 1986). This provides a basic mechanism for controlling a machine's velocity not just forward and backward but sideways as well.

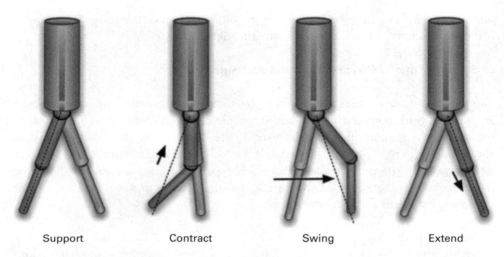

| Support | Contract | Swing | Extend |

Figure 11.1
Walking gait with jointed knees. The dotted line denotes the virtual hip angle used for foot placement.

Foot passing   To apply foot placement to a machine's gait each leg needs a way for the foot to pass the other without striking the ground. A simple artificial solution would be to use a telescoping leg. When the machine falls forward its rear leg naturally loses contact with the ground. Upon detecting this the leg contracts, swings past the other leg, and extends. When the leg finally strikes the ground the rear leg loses contact and the cycle is repeated with the other leg. If the angle of the hip is correct when the foot strikes the ground the machine can maintain a desired velocity. Not only does the hip angle need to be correct but also the leg must swing to that hip angle in just the right amount of time. This implies that any control system used to rotate the hip or knee must have good control over joint velocity. For a more human-like gait the telescoping leg model can be replaced with knees as shown in figure 11.1. It is important to note that this creates a virtual angle at the hip. On a telescoping leg the hip angle will not change as the leg is contracted, but when a knee is added, the hip angle must increase as the knee bends. A dotted line is used in figure 11.1 to denote the virtual angle that should be used when calculating foot placement.

Weight balancing   In previous work (Vaughan, Di Paolo, and Harvey 2004b) a *stance* mode (see section 11.5.2 for a detailed explanation of modes) was implemented making a positive connection between a gyroscope that detected the orientation of the waist around the x-axis and the desired hip velocity. This *stance* mode balanced a weighted torso placed above the hips. The idea of balancing weight above a machine's hips is often the focal point of research on bipedal walking. At its simplest the torso can be thought of as an inverted pendulum (Raibert 1986). Linear feedback from orientation sensors can be used to rotate the hip and balance the torso dynamically. If the torso begins to tip, the hip is rotated to capture its weight.

11.5 Stage One: Walking Forward and Backward

The purpose of stage one is to develop a planar machine similar to the one explored in previous work but built upon a more flexible control system. Movement is constrained to the sagittal plane (x- and y-axis). The degree of freedom in the hip that allows the leg to rotate to the side (around the x-axis) is locked (figure 11.2). A predesigned control system is encoded into a genotype and seeded into a population of machines that is later evolved with a GA. To simplify the problem, the use of flat spring-loaded feet is added in stage two.

11.5.1 The Body
The body used in this chapter initially has six degrees of freedom: two in each hip and one in each knee. Later in stage two ankles and feet are added, increasing the number of degrees to ten (figure 11.2). The simulation is done with the open dynamics engine

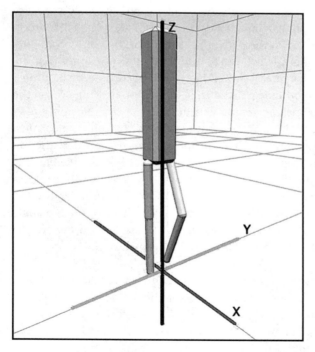

Figure 11.2
Body of walking machine with axes of movement.

(ODE) and the parameters of both the body and control system are evolved using a GA (details follow). Body parameters are evolved from the following ranges: thigh's mass (kg) [2,4], thigh's length (m) [2,4], shin's mass (kg) [1,3], shin's length (m) [0.3,0.4]. The torso's mass is 30 kg and its length is 0.7m. The range of motion for all joints on the machine was between $-\pi/2$ and $\pi/2$.

A simple muscle model is used that supports foot placement and energy conservation through passive dynamics. Each degree of freedom has three parameters: *target angle*, *desired velocity*, and *maximum torque*. The first two allow the joint to move to a target angle smoothly within a specified time as required by foot placement. The last parameter places a limitation on how much torque can be used to reach the target, giving evolution a mechanism to take advantage of the natural dynamics of the body.

11.5.2 Modes

We refer to dynamic patterns playing a functional role in walking as modes. These are implemented through individual neural circuits (similar to reflex circuits in animal walking). In previous work just two modes were used for each leg in the walking cycle: *swing* and *stance*. A winner-take-all circuit was used to switch between each mode depending on whether the foot was on or off the ground. In the more complex machine proposed here the *swing* mode could become overly complex. It must contract the leg, swing it to the proper location, and then extend it in just the right amount of time. It must keep track of the virtual hip angle and respond dynamically to changes in forward or backward velocity. The approach of this chapter is to hand-design a simple network that can be improved though evolution. However, this mode would appear to require a nontrivial, nonlinear solution. One approach is to break the *swing* mode up into several modes that have simpler solutions as done by Raibert (1986) and Pratt and Pratt (1999). Logically these are: *contract*, *swing*, and *extend*. Figure 11.3 shows the transitions between modes for each leg. Modes are implemented as networks with *sensors*, two *hidden layers*, *effectors*, and a *switching neuron* (figure 11.4).

Sensors A list of sensors and their description can be found in table 11.1. For angle sensors such as *Hip X*, *Hip Y*, and *Knee* a single sensor neuron is used. Angles are mapped linearly onto the sensor neuron's activation with negative angles corresponding to activations below 0.5 and positive angles corresponding to activations above 0.5. For velocity sensors neuron pairs are used: *forward/backward* and *left/right*. When the machine is falling forward the *forward velocity* neuron's activation increases over 0.5 while the *backward velocity* neuron's activation is maintained at 0.5. When falling backward the opposite is true. While this could have been encoded in one neuron the ability to get the velocity regardless of its sign helps to simplify the design.

Hidden layers There are two hidden layers (figure 11.4). The first hidden layer is primarily used to control movement through the sagittal plane, although there are

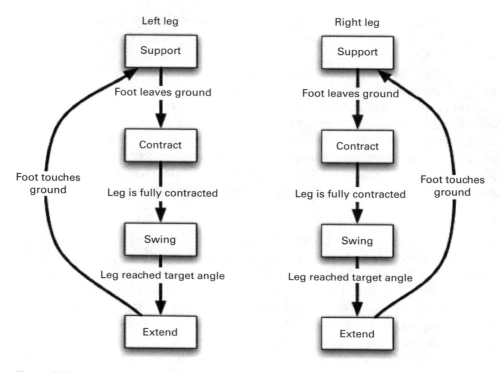

Figure 11.3
Mode diagram. Each leg has its own mode independent of the other.

some exceptions. The second layer is used in stage two to control movements laterally outside the sagittal plane.

Effectors   These are neurons that connect to each of the three powered joints: the hip around the x-axis, the hip around the y-axis, and the knee around the y-axis. Each joint is controlled by three neurons.

Tar   The activity of this neuron indicates the target angle the joint should rotate to. Its value can be anywhere between 0 and 1. If the value is 0 the joint will strive to rotate to *–range* and if it is 1 the joint will try to rotate to *+range*, where *range* is the maximum amount of movement of the joint in radians.

Vel   The activity of this neuron sets desired velocity *v* at which the joint should rotate until it reaches its desired angle. This value can be between 0 and 1. Actual velocity (*av*) is calculated as: $av = (v - 0.5)*2$.

Tq   The activity of this neuron indicates maximum torque (*mt*) that can be applied to reach the desired velocity. If *mt* is 0 the joint becomes unpowered regardless of velocity or target.

Table 11.1
The sensor neurons supplied to each mode circuit

Sensor type	Description
Hip angle X	The current angle of the hip joint as it rotates around the x-axis.
Hip angle Y	The current angle of the hip joint as it rotates around the y-axis.
Knee angle	The current angle of the knee joint as it rotates around the y-axis.
Foot contact	Becomes 1 when the foot is touching the ground, 0 otherwise.
Forward pitch	The angle of the machine's torso as it tilts forward.
Backward pitch	The angle of the machine's torso as it tilts backward.
Right roll	The angle of the machine's torso as it tilts right.
Left roll	The angle of the machine's torso as it tilts left.
Forward velocity	Velocity of the machine's torso as it moves forward.
Backward velocity	Velocity of the machine's torso as it moves backward.
Right velocity	Velocity of the machine's torso as it moves right.
Left velocity	Velocity of the machine's torso as it moves left.

Switching neuron Each mode network may connect to all incoming sensors as well as signals coming from other mode networks. The activity of the switching neuron (S) in each mode circuit can be increased or decreased by neurons in the same circuit or in others. Only one of the four modes in a leg can be active at any time so the mode with the strongest switching neuron activity is enabled and all other modes are inhibited (i.e., their effectors shut down).

Activation functions Continuous-time recurrent neural networks are used to implement the different mode networks (Beer 1995). The activation of neuron i (y_i) is given by

$$\tau_i \dot{y}_i = -y_i + \sum_j w_{ij} Act(y_j + b_j) + I_i$$

where τ_i is the decay time constant, w_{ij} is the weight of the connection between neuron j and i, b_j is a bias term, $Act()$ is the activation function, and I_i is an external input.

Five types of activation functions are used.

Type 1 $\sigma(s) = \dfrac{1}{1 + e^{-s}}$

Type 2 $\alpha(s) = |\sigma(s) - 0.5| + 0.5$

Type 3 $\eta(s) = \begin{cases} 0 & s + 0.5 < 0 \\ 1 & s + 0.5 > 1 \\ s + 0.5 & \text{otherwise} \end{cases}$

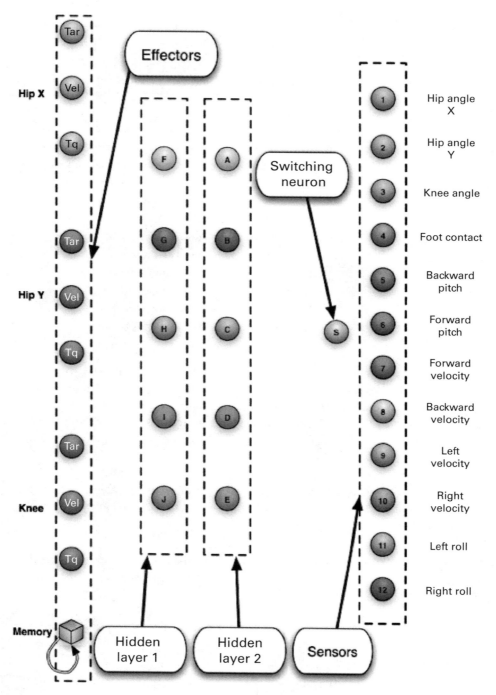

Figure 11.4
Neural network architecture for implementing each mode.

Type 4 $\beta_1(s) = \begin{cases} 1 & s \geq 0.5 \\ 0 & s < 0.5 \end{cases}$

Type 5 $\beta_2(s) = \begin{cases} 1 & s \geq 0.5 \\ 0.5 & s < 0.5 \end{cases}$

Computationally, most of these activation functions could be replaced in practice by small networks of neurons using only the sigmoid function, $\sigma(s)$. However, the use of the different functions greatly simplifies the hand-design of seed networks. As a graphic convention in the figures that follow describing neural circuits, neurons with activation function of type 1 are represented by plain circles, type 2 by squares, type 3 by cubes, type 4 by triangles, and type 5 by diamonds. Excitatory connections are represented by full lines, inhibitory connections by dashed lines, bias neurons by double circles, and modulatory connections as arrows linking a neuron and another connection.

11.5.3 Support

The *support* mode becomes active when the foot is touching the ground. It attempts to balance the upper torso of the biped by rotating the hip either forward or backward. A simple algorithm is:

• Keep the knee straight.
• If the torso is pitching forward power the leg backward until the torso becomes upright. If the torso is pitching backward power the leg forward until the torso becomes upright.

The support network (see figure 11.5) has four tasks: keeping the knee straight, keeping the torso upright, storing the angle of the leg, and ensuring that when the leg leaves the ground the *contract* mode is triggered.

Keeping the knee straight The knee is kept straight by giving the (*B*) neuron a positive bias and excitatory connections to the knee angle target, velocity, and torque. This causes the knee to gently push into the kneecap.

Keeping torso upright Connections from the forward (6) and backward (5) pitch sensors increase the hip's velocity and torque. The more the torso falls forward or backward the more speed and torque can be used to bring it upright again. A positive connection from (5) and a negative connection from (6) set the hip's desired target angle toward the direction the leg needs to rotate.

Storing the hip angle Why store the hip angle? Once the leg loses traction with the ground the leg switches to the *contract* mode. In this state the leg must support itself at the angle at which it left the ground and retract. This requires some way to store the last hip location. A single positive recurrent self-connection on neuron (*M*) is used to store the current angle by adding a negative and positive connection to neuron (*C*). Whatever value (*C*) receives will automatically be stored in neuron (*N*).

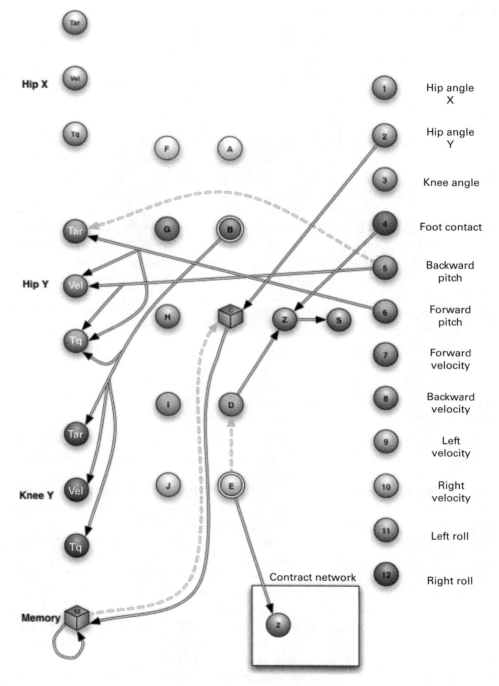

Figure 11.5
Network circuit design for the *support* mode.

Triggering the contract mode   Once a foot loses traction its touch sensor inhibits the switching neuron and then *contract*, *swing*, and *extend* modes have to compete to see which mode gains control. Each one needs to be active at different stages of the walking gait. One solution is to train the connections from a mode's input layer to its switching neuron such that it becomes active at the right time. However, these functions may not be linear, requiring an additional hidden layer between the inputs and the switching neuron. A simpler approach is to notice that the *support* mode always precedes the *contract* mode. It is much easier to allow the *support* mode to tip the vote through a positive connection to *contract's* switching neuron. Neuron (*E*) is given a positive bias and connected to the *contract* network's (*Z*) neuron. The (*Z*) neuron has a larger time constant (τ) than other neurons so while the *support* mode is active, it charges up. When *support* does finally lose, (*Z*) is still excited momentarily, resulting in a higher activity in *contract's* switch neuron.

11.5.4 Contract

The *contract* mode's task is to contract the leg while maintaining the current hip angle. When the leg is fully contracted to the desired leg height it triggers the *swing* mode. The *contract* network is shown in figure 11.6.

Contraction speed   One important factor in walking is the speed at which the leg moves. This is not only how fast the leg swings forward but also how fast it is contracted, swung, and extended. The total time it takes for all three to occur is critical to good foot placement. Too slow and the leg will not reach the target in time and the machine will stumble. Too fast and unnecessary energy is used that could have been saved. The speed of leg contraction should be proportional to the speed the machine is moving forward or backward. This is the function of neuron (*C*), which is excited by the forward (*7*) and backward (*8*) velocity sensors. It in turn excites the knee velocity/torque and the hip velocity/torque. The faster the machine falls the greater the strength and velocity of contraction.

Contraction height   Contraction height is specified by neuron (*D*) whose positive bias pulls the knee up and the foot in at the velocity specified by neuron (*C*). Future stages could use this neuron to increase leg height on more rugged terrain.

Virtual hip angle   The leg must contract but keep whatever angle it had when it was in the *support* mode. Neuron (*B*) adjusts the leg angle by adding the last angle stored in *support's* (*M*) neuron.

Triggering swing   The *swing* mode is triggered by thresholded neuron (*E*) when the leg is fully contracted. Neuron (*E*) receives the difference between the current knee angle (*3*) and the desired contraction height (*D*). When the sum of the two is greater than 0, neuron (*E*) reaches its threshold and fires, inhibiting *contract* and exciting the *swing* mode.

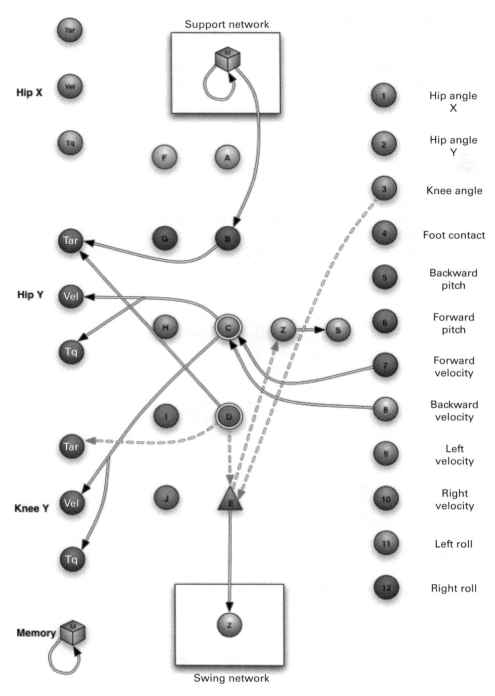

Figure 11.6
Network circuit design for the *contract* mode.

11.5.5 Swing

The purpose of the *swing* mode is to generate proper foot placement and move the leg such that when extended it will break the machine's fall and reduce its speed. Once the leg has reached its position it triggers the *extend* mode. The *swing* network is shown in figure 11.7.

As discussed earlier, the speed at which a leg moves is critical for proper foot placement. The leg must swing forward or backward proportionally to the machine's velocity. The faster it falls the faster the leg needs to move to catch the machine's weight. Foot placement is also proportional to the velocity of the machine. Both of these mechanisms together produce a balanced walking gait.

The swing algorithm uses six neurons: (A), (B), (C), (D), (E), (J), and (M).

Neuron (A) stores the virtual hip angle (factor three) in (M) by subtracting the desired contraction height (D) from the actual hip angle (2).

Neuron (B) computes the velocity and torque of the hip as it swings forward or backward. Factor one is implemented by positive and negative connections from forward (7) and backward (8) velocities. Factor two is implemented by adding in the previous leg position stored by the *support* mode.

The knee is kept bent at its contraction height due to the lack of any connections to its velocity combined with (C) constantly exciting the torque.

Neuron (D) is proportional to the desired contraction height. It has connections to the hip target to keep the leg contracted throughout the swing phase.

Neuron (E) computes the absolute error between the target hip angle and the actual hip angle. A large error keeps (J) below its threshold; when the error is small enough (J) fires and triggers the *extend* mode.

In previous work (Vaughan, Di Paolo, and Harvey 2004a) a machine was evolved that could be controlled easily by gently pushing it at the speed and direction required. If no force were applied the machine would stand still. Such control mechanisms are very desirable when building machines for carrying loads, as they do not require a driver. Instead they can be gently pushed or pulled by a rope in the desired direction. If a more complex control system is needed it can be added incrementally later. In this chapter, we explore a subsumptive control of such a mechanism by adding a special neuron (K) to modify the default behavior. As discussed earlier, foot placement can be used to directly control the velocity of a walker. Too large a step and the walker will slow down; too small a step and the machine will continue to accelerate as it falls. Neuron (K) can adjust the machine's velocity simply by connecting it to the *Hip Y* target neuron. As (K) is increased the machine tends to walk forward as it fails to catch its center of mass (COM) in time. If gently pushed backward it will overcompensate and take too big a step backward causing it eventually to walk forward again. When (K) is inhibited the opposite happens and the walker tends to walk backward. The actual speed of the walking gait depends on (K) allowing basic control over acceleration and deceleration of the machine.

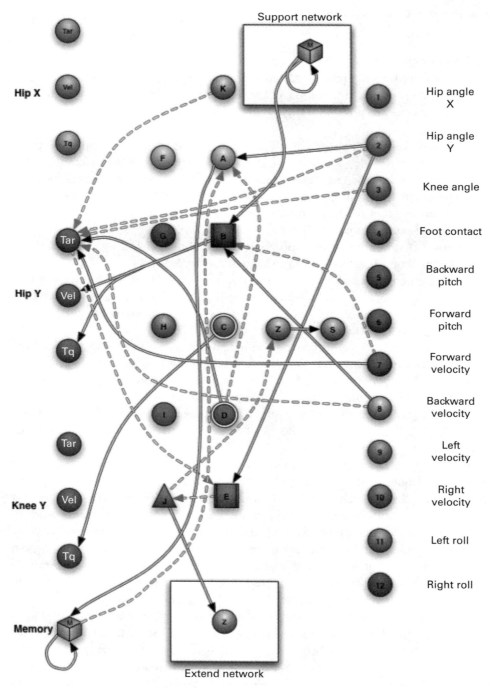

Figure 11.7
Network circuit design for the *swing* mode.

11.5.6 Extend

The *extend* network's job is the opposite of the *contract* network. It must straighten the leg while keeping it fixed at the desired angle stored by the *swing* network. Once the foot touches the ground the *support* network is automatically triggered.

Only two neurons are required: (*C*) and (*B*) (figure 11.8). Neuron (*C*) has a bias and supplies a constant velocity and torque to the knee and hip to keep them firm when the machine is not moving. When the machine has velocity it is added to (*C*). The faster the machine falls the faster and stronger the leg extends out to catch its fall.

11.5.7 Experiments in Flat Surfaces

A geographically distributed GA (Husbands 1994) was used with twenty-five individuals. Each generation, 20 percent of the genes were selected according to fitness and changed using creep mutation to populate the next generation. The mutation rate was 0.02.

Machines were evaluated by testing their ability to walk both forward and backward. Two test cases were used. In the first test case the (*K*) neurons in each *swing* leg were set to a random negative value causing the machine to walk backward. In the second case a random positive value was used, causing the machine to walk forward. Five trials were used for each test case and then the averages for each case were multiplied together.

The following fitness function was used:

$$fitness = time * min(rot) * min(vel) * min(energy)$$

where $min(t) = 1/(1 + t)$, *time* is the amount of time the machine walked (maximum is 30 seconds), *rot* is the absolute average error between 0 and the body's pitch angle, *vel* is the absolute average error between the desired velocity—specified by the (*K*) neuron—and the actual velocity, and *energy* is the average torque used by all actuators.

An evaluation was started by placing both legs on the ground and stimulating the switching neuron for *support* on one leg while stimulating *contract* on the other. Once a leg lost contact with the ground all stimulation was removed.

11.5.8 Observations and Improvements

The hand-designed networks did well even before any evolutionary processes were applied with an average fitness of 0.4. After evolving for 1,320 generations it improved to a fitness of 0.8.

The machine walked well on a flat surface but in real-world environments this is rarely the case. On less smooth surfaces the machine's natural walking dynamics must be able to minimize disturbances in the torso in the same way shock absorbers work

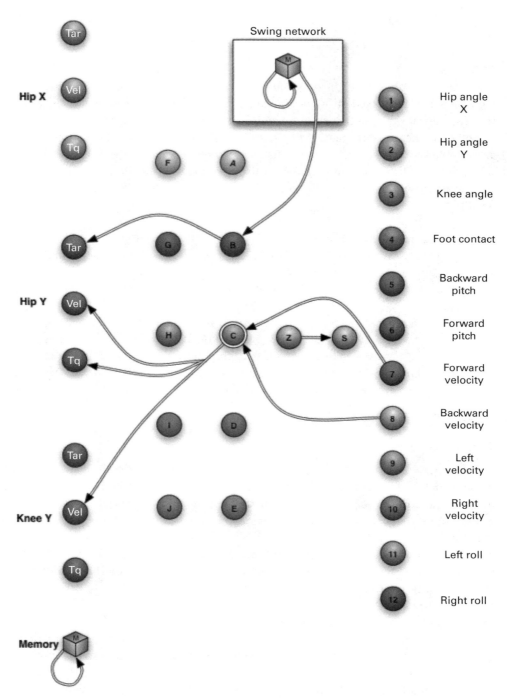

Figure 11.8
Network circuit design for the *extend* mode.

for a car. A bumpy ground surface was simulated using a mesh of flat triangles. The mesh was a grid of 70×70 squares each split into two triangles with random heights. The maximum height of the surface was 3 cm. The same GA population evolved on the flat surface was tested on the rugged surface. Initially their average fitness fell from 0.8 to 0.12. After 400 generations the machine's best average fitness was only 0.24.

Although the machine's performance is excellent on flat surfaces, when placed on bumpy, uneven surfaces the machine destabilizes and loses its balance, falling either forward or backward. Through careful observation the cause for this poor performance was isolated to four different situations. The first two were early foot strike when stepping on a small incline and late foot strike when stepping into a small depression. The third situation was when the machine lost forward momentum and came to a complete stop with both legs side by side. Eventually, it began to fall either forward or backward while both feet continued to stay on the ground. There was no triangle between each foot and the torso so neither leg could be lifted off the ground as the torso moved. Unable to switch to the *contract* mode the walking gait stopped completely and the machine fell. The fourth situation was when both legs accidentally lost traction with the ground. In this situation both legs would enter the *contract* mode simultaneously and the machine would fall to its knees.

In the early foot strike situation, the machine steps onto an incline and the leg strikes the ground before it is fully extended. When the leg does finally extend, it is during the *support* mode, which causes the machine to push itself backward (figure 11.9). The simplest remedy for this behavior is to modify *support* to reduce the velocity at which it can straighten the leg.

In late foot strike, the foot extends fully but does not strike immediately due to a small depression on the ground. As the foot moves into the depression the machine's center of mass falls too far forward, causing it to fall (figure 11.10). The easiest remedy is to wait until the leg is fully extended, then if the foot still hasn't struck the ground, bend the knee on the opposite leg. This will push the extended leg farther down until it strikes the ground, preventing the machine's mass from falling too far forward. To modify the modes, knee angle (3) causes neuron (A) to fire when it reaches full extension. Neuron (A) in turn excites neuron (F), which due to its longer time constant fires a few moments later. Neuron (F) triggers the knee neurons in the opposite leg's network to contract momentarily.

Correcting the third situation when the walking gait is stalled as both legs come together requires a different approach. This is a fundamental problem in walking. What happens when a person is standing and they suddenly begin to fall backward? They naturally lean to one side, pick up the opposite leg, and take a step back. This can be looked at as an additional mode called *stand* whose purpose is to contract the opposing leg if the machine begins to fall while its legs are together (figure 11.11).

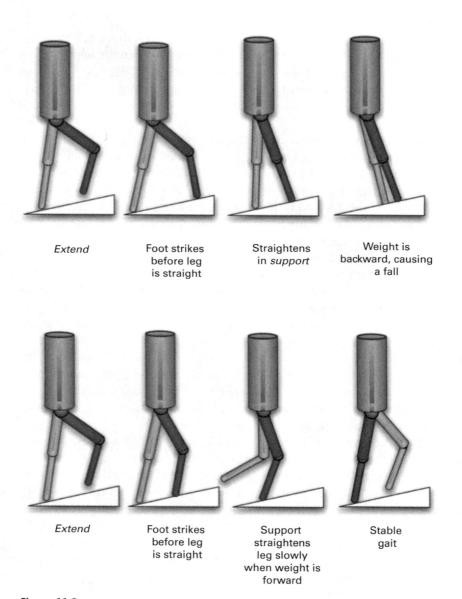

Extend

Foot strikes
before leg
is straight

Straightens
in *support*

Weight is
backward, causing
a fall

Extend

Foot strikes
before leg
is straight

Support
straightens
leg slowly
when weight is
forward

Stable
gait

Figure 11.9
Early foot strike (top), strategy for overcoming early foot strike (bottom).

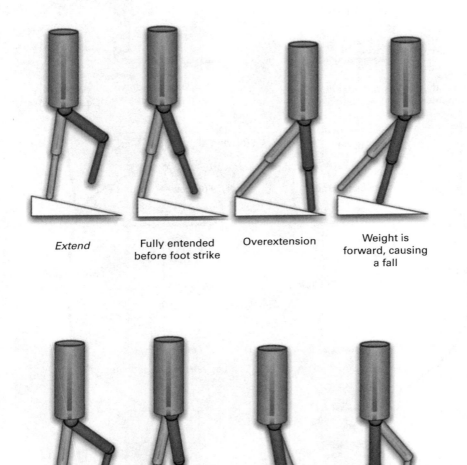

Figure 11.10
Late foot strike (top), strategy for overcoming late foot strike (bottom).

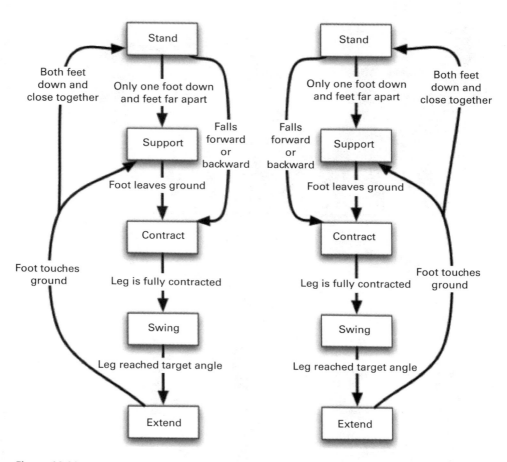

Figure 11.11
Addition of the *stand* mode.

This situation also happens at the beginning of our experiments when the machine is first placed on the ground and pushed.

Currently this is solved artificially by placing one leg in *support* mode and one in *contract* mode at the very beginning of the experiment. This can easily be replaced by a *stand* mode allowing the machine to initiate the first step itself.

11.5.8 Stand

The *stand* mode inherits all the functionality of the *support* mode while adding some additional features. Five additional neurons are used: (Y), (G), (H), (I), and (J) (figure 11.12).

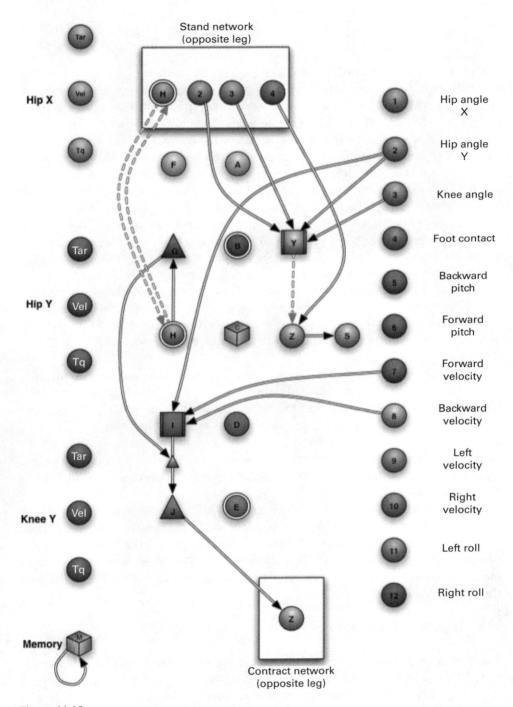

Figure 11.12
Network design for the *stand* mode. This mode inherits all the connections of the *support* network while adding functionality. Inherited connections are not shown.

The (*Y*) neuron responds to the difference between each leg's virtual angle. The closer the legs are to each other the smaller the virtual angle becomes. In turn, it inhibits (*Z*), which stimulates the switching neuron. The result is that the legs must be very close to each other in order for the *stand* mode to become active.

Neuron (*I*) is excited if the machine begins to fall forward or backward; if it falls too much (*J*) will fire contracting the opposite leg's knee. As both legs are in *stand* mode only one (*I*)–(*J*) circuit should activate or both legs will simultaneously contract and the machine will fall. Normally this would be determined by weight. If the center of mass (COM) is over the left foot it is natural to contract the right one and vice versa. However, the machine in stage one is planar and is supported laterally such that the COM is equally distributed between each leg. In stage two when lateral support is added, the hip angle X can be used to decide which leg the COM is over. To remedy this in planar experiments, the concept of *handedness* must be introduced. A small random initial bias is given to each (*H*) neuron so when the *stand* mode is entered the neuron with the greatest bias will become active.

Neuron (*G*) has a threshold causing it to fire only if the network's (*H*) is the winner of the competition. It acts as a simple switch turning the connection between (*I*) and (*J*) on and off.

As discussed earlier, the fourth situation that destabilizes the machine on irregular surfaces is when both legs lose traction simultaneously. When this happens it is important that one of the legs remain in the *support* mode until the machine touches the ground again. An easy solution is to create positive lateral connections from *contract*, *swing*, or *extend* on one leg to the *support*'s switching neuron on the other. The *support*'s switching neurons should be more excited when a mode other than *support* on the other leg is active. A specialized neuron with a threshold activation function (type 5) is used that only fires if the mode is active.

Overall performance improvement The four modifications were made and a new population of twenty-five machines was evolved under the bumpy surface condition. After the first generation the population's average fitness was 0.2, very close to the best average of the unmodified population of 0.24. The fitness quickly grew until machines appeared that could walk the entire evaluation period. After 260 generations the average began to flatten out at 0.54.

11.6 Stage Two: Lateral Control

The goal of stage one was to develop a planar machine that could walk on irregular surfaces. In this second stage the virtual boom is removed and the machine is allowed to walk unsupported. The machine must now work to keep its torso from falling or rolling to the left or right outside of the sagittal plane.

Two basic walking principles can greatly simplify understanding lateral balance: lateral foot placement and weight shifting. Lateral foot placement is fundamentally the same concept used in the sagittal plane. If the machine detects it is falling to the right it must also move its foot to the right to catch its weight as it falls and vice versa.

If a person is standing and wishes to take a step with her right foot she must first have her COM positioned over her left foot. If she doesn't, she will have to step out to the right to stop her lateral momentum. This weight shift allows a person to step in a straight line forward. This can be done mechanically with a damped spring in the hips. For example, if the machine is taking a step forward with its right foot the COM will begin to move toward the right foot. When the right leg eventually touches the ground the COM compresses into the spring and balances over the right hip. When the left leg contracts this potential energy is released and the COM is pushed back toward the left hip. This creates a kind of throw-and-catch game with the COM between the legs.

11.6.1 Support

Lateral support along the y-axis addresses a problem similar to that of the sagittal plane; it must keep the torso upright as the machine begins to fall. This can be accomplished by four additional neurons: (F), (A), (G), and (T) (figure 11.13). The (F) neuron takes input from the roll neurons (11, 12) and adjusts the target lateral hip angle along the x-axis to keep the torso upright. Torque and velocity are controlled by (A), which takes the absolute value of (F).

Unlike support moving along the sagittal plane, support along the y-axis must also control the shifting of the COM from one leg to the other. At its simplest, if the *support* mode detects that the opposite leg is not in *support* (contracting) it should begin to push its leg to the outside moving the COM over the contracting leg. When the opposite leg does finally touch the ground the COM will now be resting above it. This can be done with just two neurons (G) and (T). Neuron (T) has a threshold and fires only when the *support* mode is active. It in turn inhibits the (G) neuron in the opposite leg. Neuron (G) has a small bias and an inhibitory connection to (F). If both (T)s are firing then both legs are down weakening the bias of both (G)s. In this case (G) has no effect on the hip angle. If the opposite leg moves into *contract* its (T) will stop firing and (G)'s bias will cause it to excite (F) moving the leg to the outside. The result is that the COM begins to move over the newly contracting leg.

When a leg enters the *support* mode the COM will be over it. The task of this mode is to capture this mass and then release it when the opposite leg contracts. This can be accomplished with a mechanical spring (Vaughan, Di Paolo, and Harvey 2004a). Catching of the COM only happens as the COM moves from the inside leg toward

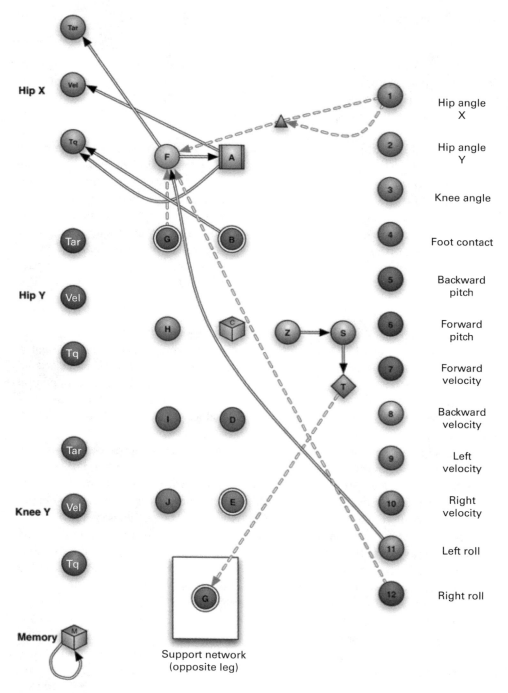

Figure 11.13
The *support* network for lateral movement outside the sagittal plane.

the outside, requiring this spring to be enabled only when the hip angle X (*1*) is negative. A spring can be approximated by making an inhibitory connection from the hip angle X (*1*) to the (*F*) neuron. An additional inhibitory connection disables this connection when (*1*) is a negative angle. This creates a spring that only engages when the COM is falling toward the outside of the leg. It is up to the other leg to capture the COM if it falls to the inside of the leg.

11.6.2   Contract

The task of the *contract* network is to keep the leg laterally stable while the leg is lifted. If the leg is angled out to the left or right, the *contract* network slowly brings the leg back toward the center. Ideally the leg should be angled slightly underneath the torso, reducing the distance the COM needs to be shifted on each step.

A single neuron (*I*) is used to provide a small bias that sets the desired hip angle, velocity, and torque (figure 11.14). Its connection to the hip's velocity is relatively small so the machine doesn't jerk its leg but moves it slowly under the hip.

11.6.3   Swing and Extend

During the swing phase it is possible for the machine to begin to fall to the left or the right. Modifications must be made to reduce velocities along the y-axis when this happens. A single neuron (*H*) is used to track the error between the desired lateral velocity (always 0) and the current lateral velocity (figure 11.15). Neuron (*H*) receives inputs from the right (*10*) and left (*9*) velocity sensors and though (*G*), adjusts the desired hip target, velocity, and torque. The (*G*) neuron from the opposite leg's *support* network provides critical error correction information. Neuron (*G*) as discussed earlier in the *support* network, becomes excited when the COM shifts over the opposite leg. If (*G*) is active then the machine will be falling toward the opposite leg at a velocity proportional to (*G*). By adding this information to the (*H*) neuron we keep the leg from using foot placement to correct the shift velocity. This lateral velocity is a natural part of the gait and should not cause the legs to compensate for it. The *extend* network is identical to that of *swing*.

11.6.4   Stand

The *stand* mode becomes active when both legs are together and both feet are touching the ground. At this point the COM should be over one of its feet in case it needs to take a step in the future. In the planar machine the (*H*) neuron competed with the other leg's (*H*) neuron to decide which leg would be in *support* mode and which would become *contract*. In this stage the (*H*) neurons can receive input from the hip angle X (*1*) forcing the leg that is supporting more of the COM to win. This winning state gently shifts the COM completely over its leg by slowly bending its knee. To keep the opposite leg relaxed during the weight shift (*G*) disables the connection between (*F*)

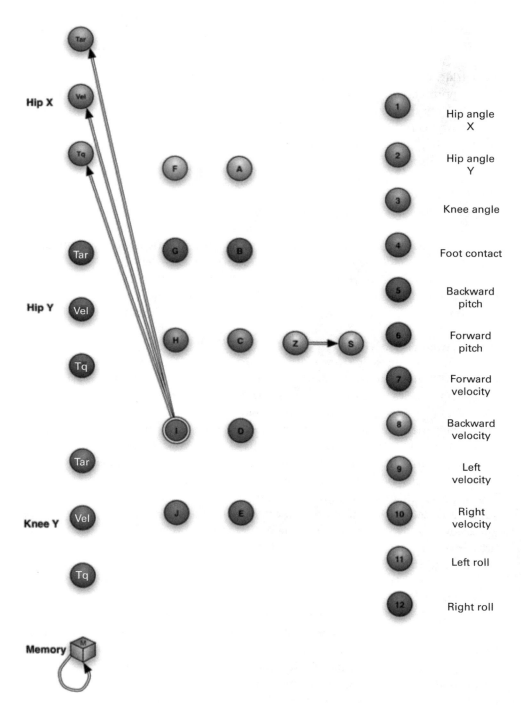

Figure 11.14
The *contract* network for lateral support.

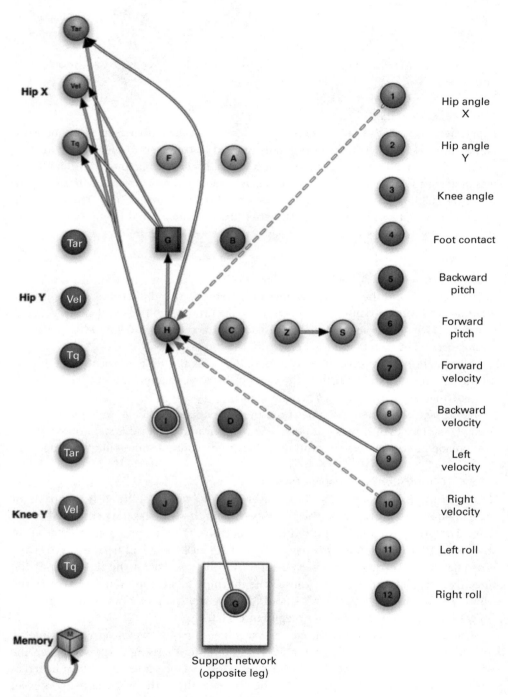

Hip X

Hip Y

Knee Y

Memory

Support network
(opposite leg)

1 Hip angle
X

2 Hip angle
Y

3 Knee angle

4 Foot contact

5 Backward
pitch

6 Forward
pitch

7 Forward
velocity

8 Backward
velocity

9 Left
velocity

10 Right
velocity

11 Left roll

12 Right roll

Figure 11.15
The *swing* network for lateral movement outside the sagittal plane.

and (*A*) when it is not firing. The resulting machine always keeps its weight over one of its feet.

11.6.5 Experiments

Experiments on a flat surface were conducted in the same manner as in stage one with a few minor changes. The forces normally applied to the torso to prevent any movement other than along the sagittal plane (a virtual boom) were removed. An addition was made to the fitness function to minimize lateral movement from side to side. This improves stability by selecting individuals who shift their weight only when needed. The mutation rate was lowered to 0.01 after observation showed lateral stability was more sensitive to these kinds of changes. Fitness is defined as follows:

fitness = time * min(rot) * min(vel) * min(energy) * min(lvel)

where $min(t) = 1/(1 + t)$, *time* is the amount of time the machine walked (maximum is 30 seconds), *rot* is the absolute average error between 0 and pitch/roll/yaw angles of the torso, *vel* is the absolute average error between the desired velocity and the actual velocity, *energy* is the average torque used by all actuators, and *lvel* is the absolute average error between 0 and the velocity along the y-axis.

A population was evolved for 190 generations on a flat surface. The average fitness in the first generation was 0.16 but over the next 80 generations rose to 0.6 and then flattened out over the next 110 generations.

As in stage one, when machines from the flat population above were placed on an irregular surface their fitness was poor, averaging around 0.1. In this stage observations were made directly before attempting to continue evolution on a bumpy surface.

11.6.6 Observations and Improvements

If a machine is built using just foot placement and weight shifting, it is capable of walking quite well on a flat surface. However, as in the first stage, when placed on an irregular surface the machine often lost its balance. One reason could be that all of the innovations to handle *early foot strike* and *late foot strike* had been evolved while on a flat surface. The other observed cause was *foot tangling*. If the right foot of the machine stepped into a gully it caused the machine to shift too much weight to the right. To compensate the machine tried to move the left foot toward the right only to get it tangled up with the right leg (figure 11.16).

The easiest way to modify the current machine is first to develop a small symmetric neural network that inhibits or disables a neuron if foot passing is allowed, given the virtual leg angles and forward and backward velocity. Once computed it can be fed as an input to any state that requires it. The center of the entire walking network is an ideal location for this new leg-crossing network as it has access to all state sensors, allowing a new sensor to be added (*13*). The basic method to keep the legs from getting tangled is as follows:

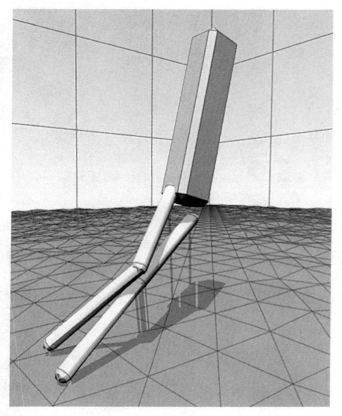

Figure 11.16
Foot tangling: while walking on an irregular surface the machine begins to fall to its left. To compensate it swings its right leg directly into the left, knocking its hip out of its socket.

1. *Side = left* or *right*.
2. If the machine is moving forward while falling toward the *Side* and the *Side* leg is in front of the opposite leg, disable the opposite leg's ability to move toward the *Side*.
3. If the machine is moving backward while falling to the *Side* and the *Side* leg is in back of the opposite leg, disable the opposite leg's ability to move toward the *Side*.

11.6.7   Foot Tangle Network

The neural controller for foot tangle can be seen in (figure 11.17). To determine whether the machine can move a leg laterally the following information must be considered:

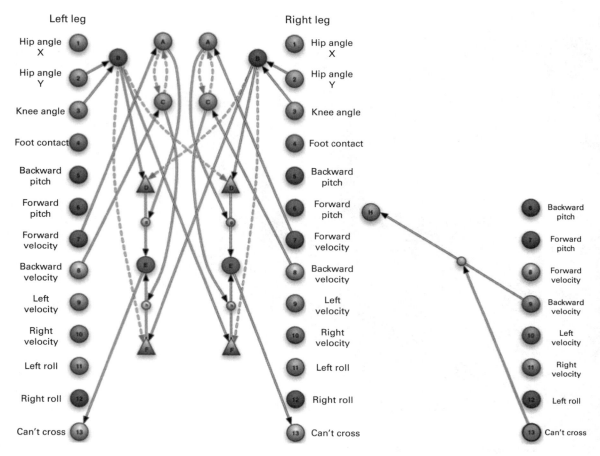

Figure 11.17
Implementation of the *tangle* network (left); modifications to the *swing* and *extend* networks (right).

1. What direction is the machine walking? Forward or backward?
2. What are the virtual leg angles?
3. What is the difference between these angles?

Neurons (*A*) and (*C*) create a winner-take-all circuit to determine which way the machine is walking. Neuron (*A*) takes input from the forward velocity sensor and (*C*) from the backward one. Due to slow time constants (*A*) and (*C*) are resistant to quick fluctuations in velocity. If (*A*) becomes active the machine is moving forward, if (*C*), backward. Neuron (*B*) estimates the virtual angle of each leg and (*D*) and (*F*) calculate

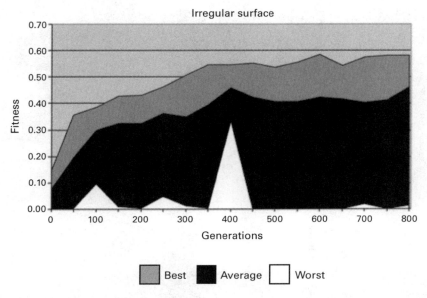

Figure 11.18
Fitness after modifications to prevent foot tangling.

the difference between the leg angles and fire when one leg is in front of the other. Neuron (*D*) triggers (*E*) when walking backward and crossing is not allowed. Neuron (*F*) triggers (*E*) when walking forward and crossing is not allowed. An additional sensor (*13*) is used to propagate this information to each mode network.

On the *swing* and *extend* modes connections are made that disable the connections between the left velocity sensor when crossing is not possible (figure 11.18, right). Due to the bilateral nature of the network the left and right sensors are swapped on the left side of the body. As a result each leg is prevented from moving inward under the body where it could get tangled up with the other leg when sensor (*13*) has a high activation.

Figure 11.18 shows the improvement in fitness after including the foot tangle network in the population seeded into the GA. When the best machine is observed, it shows quick movements when necessary both forward and laterally in response to environment disturbance. Figure 11.19 shows it quickly stepping in front of its right leg to try and capture the COM. Figure 11.20 illustrates how the machine can adjust its stride dynamically when moving over rugged terrain.

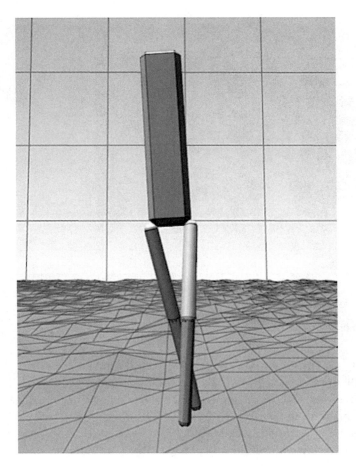

Figure 11.19
A machine stepping in front of its right leg to try to capture its COM.

11.7   Spines, Hips, Arms, and Toes

It is possible to build more complex walking machines using the techniques explored in this chapter. Preliminary experiments were done with jointed spines, rotational hip joints, arms, and toes. Each new control system was added on top of earlier successful ones. To add a spine, the body orientation sensors were moved to the head and a simple control system was used to balance the spine on top of the legs. Arms were added that passively swing from side to side on each step. An extra degree of freedom allowed the hip joint to rotate around the z-axis and a spring-loaded toe increased traction. Even with these radical changes evolution managed to integrate them into a

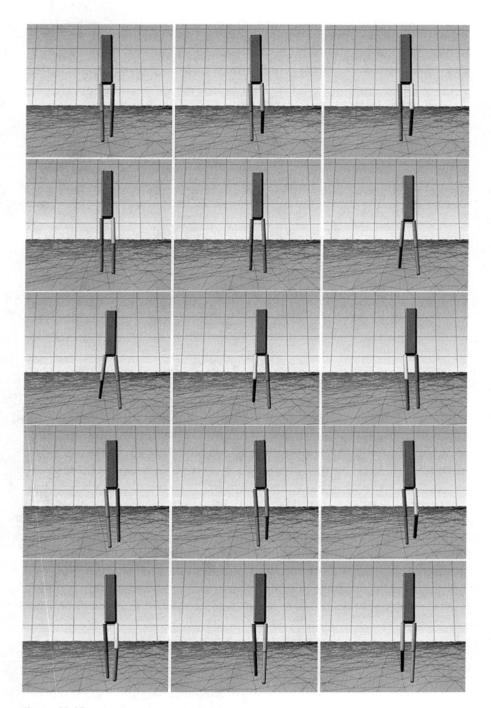

Figure 11.20
Front view of machine recovering as it steps in a gully on 5 cm surface.

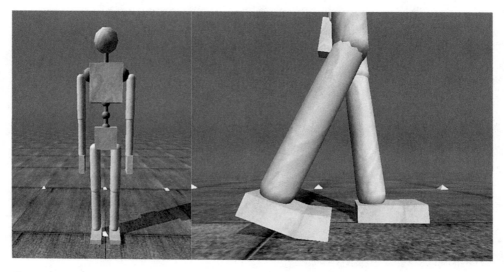

Figure 11.21
Preliminary walker with spine, head, ankles, and arms. The machine had thirty-five degrees of freedom: two in each ankle, one in each knee, three in each hip, nine in the spine, three in each shoulder, one in each elbow, two in each wrist, and one in each toe. The toe joint used a simple linear spring that provided additional traction as the heel lifted off the ground. This tended to prevent the body from twisting as it took each step.

natural walking gait. Figure 11.21 shows a machine with thirty-five degrees of freedom that can walk indefinitely. While this machine was constrained to a flat surface due to limitation of the simulation, it does show that such incremental methods can scale up to similar complexities currently explored by trajectory-based machines. This machine was called "Spine Walker."

The design of Spine Walker is based on ideas proposed by Raibert (1986) and Pratt and Pratt (1999). Like Raibert and Pratt's work this machine was built using hand-wired dynamic equations instead of neural circuits. When both approaches were compared, the neural network-based approach used throughout this chapter appeared to be more evolvable. Although the more complex spine-based machine did not use the neural circuits previously described, it still was built using SED and is another good illustration of our methodology. In this section we discuss some of the ideas revealed by our work on the Spine Walker and how they could be transferred to the neural network-based approach presented in earlier sections.

The Spine Walker was developed in five main stages:

Stage 1. The machine was evolved to walk on a flat surface.
Stage 2. An extra degree of freedom was added to the hip.

Stage 3. A flexible toe joint was added.
Stage 4. A flexible spine was added.
Stage 5. A head and arms were added.

Hips with three degrees of freedom Spine Walker has one more degree of freedom in the hip than the other machines in this chapter. This allows it to rotate the hip around the z-axis the way humans do. The control system for this is quite simple. In the *swing* mode the opposite leg's hip rotates around the z-axis to bring the hip farther forward (figure 11.22). This results in a more natural-looking walk. In terms of the current circuit model, this could be done by creating connections from the *swing* mode of one leg to the *support* mode of the other. When the leg lifts off the ground and enters *swing* it can rotate the opposite leg's hip forward.

Flexible toe joint A spring-loaded toe joint is added to each foot to make the machine walk with a more human gait. This change was observed to increase stability by increasing traction with the ground as the heel was lifting. The increased traction reduced the chance of the body twisting (yaw) too much and losing its balance (figure 11.22, right).

Spine with nine degrees or freedom The second stage is to add a spine on top of the hips that support a weighted torso. The goal of the spine is to use an inclinometer in the torso to balance the torso on top of the hips. It uses a balancing algorithm similar to the one used by the *support* mode earlier in this chapter. Its only input is that of the inclinometer and its output the direction the spine should bend around three axes to support it. The control system bends the spine to isolate movements in the hips so they do not affect the torso. This idea was taken from the observation that when people walk their upper torso tends to be relatively still compared to their hips. Once the spine is in place the population was evolved until the machine could walk without falling (figure 11.22).

Head and arms The final stage is to add a head and passive arms to the machine. This proved relatively trivial. The arms are damped and allowed to swing freely. When a leg enters the *swing* mode it causes the shoulder on the opposite arm to swing forward slightly leading to a natural-looking smooth gait. This could be done in our neural model by creating connections between the *swing* network and motor neurons in the opposite arm (figure 11.23).

11.8 Conclusions

Criticism leveled against ER has often invoked the difficulty of breaking the complexity barriers when the design process is approached from scratch. We have demonstrated here that it is indeed convenient to use evolutionary methods iteratively in combination with domain knowledge and analysis of intermediate results. This can help

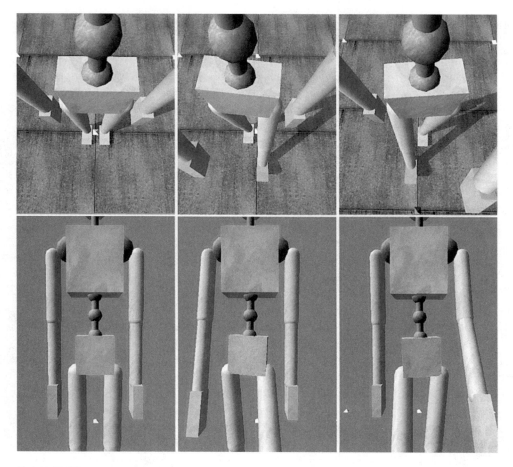

Figure 11.22

Hip movement of Spine Walker. The upper three frames show how the hips can rotate around the z-axis due to the addition of an extra degree of freedom. The lower three frames show how the hips can move from side to side due to a gentle flexing of the spine.

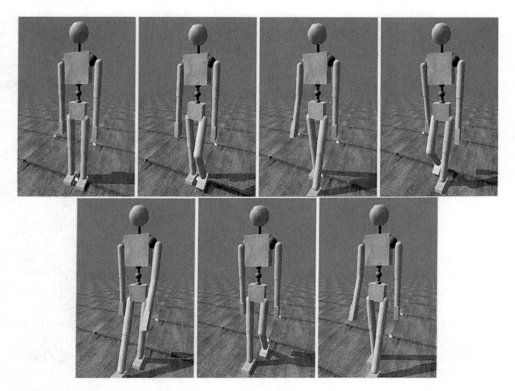

Figure 11.23
Spine Walker walking on a flat surface.

"orient" the process in directions that may be originally unknown to the engineer but that are "naturally" suggested by the process itself. In doing this, not only do we break complexity barriers for our particular design problem, but we can also draw some concrete lessons applicable in wider domains.

The machine developed in this chapter was built upon experience gained from previous work combining evolutionary design of body and controllers with engineering insights about the problem domain. The basic elements of walking were reexamined and simplified into stages involving foot placement, foot passing, and torso balancing. In foot placement simple linear relationships were shown between velocity and hip angle. To support foot passing the walking gait was broken down into five networks: *stand*, *support*, *contract*, *swing*, and *extend*. Each mode was hand-designed using simple neural circuits. A larger network was constructed that could modify these modes allowing only one active circuit per leg to be active at any given time. Through contraction and extension each foot could pass in front of the other. The *support* mode allowed the leg whose foot was touching the ground to dynamically support a weighted torso. If the torso began to tilt the hip was powered to capture its center of mass. The wider lesson of this process is that whenever feasible, a complex behavior should be approached as composed of multiple elements which may be initially designed separately using existing knowledge and then integrated by seeding these elements into the evolutionary algorithm and improving them incrementally.

In this chapter, the process of incremental design though seeding networks and stages continued to produce integrated networks with increasingly complex behavior. This follows a practical rule of thumb: increment task complexity or environmental difficulty in stages. In stage one the machine was first tested on a flat surface and then on an irregular one. Though observation four situations were found that reduced its fitness on rugged surfaces: early foot strike, late foot strike, loss of momentum when feet come together, and when both feet lost traction with the ground completely. The analysis of these issues led to network modifications to remove the problems. In stage two the virtual boom that constrained the machine to the sagittal plane was removed. This required modification of the five modes. As in stage one, the machine was tested on both flat and rugged surfaces. On flat surfaces the machine could walk quite well even before being evolved. On rugged surfaces the machine tended to tangle its legs together when trying to regain lateral balance. A foot tangle network was designed that inhibited leg movement that might cause the legs to collide, increasing fitness. This suggests that ankles are not necessary for basic walking but they can prevent twisting and inject energy into the gait. When walking normally the leg movements were slow and smooth but when destabilized due to a rugged surface the machine made quick movements to recover.

Preliminary work explores more complex machines including a jointed spine, arms, and extra hip joints. As before, each stage is built on top of earlier ones by adding simple control systems that evolution integrated into the whole. The result is a machine with thirty-five degrees of freedom that can walk infinitely on a flat surface.

Overall these machines show the power of incremental design using evolution. They can walk in any direction on smooth and rugged surfaces. They can walk slowly at 0.1 *m/s* or fast at 0.3 *m/s*. They can be scaled up to control many degrees of freedom mimicking human-like movement using spines and arms. These machines suggest that passive dynamic walkers can be scaled up to a level of human-like complexity that until recently was only explored by trajectory-based approaches. The PDWs exhibit natural walking gaits in 3D space and can switch behaviors depending on the state of their environment. Like trajectory-based machines, when pushed they can recapture their center of mass even on rugged surfaces in a controlled way and in the presence of a large number of degrees of freedom. However, these machines have the advantage of greater efficiency, speed, and simplicity. The control systems are relatively simple and could be implemented with cheap embedded controllers. They are fast, efficient, robust, and can perform different behaviors making them more practical for legged vehicles and lifelike toys.

To build a physical machine based on these simulations two issues must be addressed: the need for low impedance actuators, and any mismatches that arise between reality and simulation. There are solutions for the first issue, such as Series Elastic Actuators developed at MIT by Robinson et al. (1999), or the Programmable Spring developed at Sussex by Bigge (2010). For the second issue, there are known methods for "crossing the reality gap" (Jakobi 1998), and there are indications in this chapter, mentioned earlier, that at some new incremental stages walking was achieved even before further evolution, that the outcome of this design process is surprisingly robust, and hence more likely to be robust enough to cross the reality gap.

The techniques used in our case study have achieved results beyond the former state of the art. The design insights discussed here are specific to bipedal walking machines, but it is hoped that this case study may also provide useful insights and lessons for people trying to use an incremental ER methodology in other domains, especially in connection with the iterative strategies for generating, combining, and seeding into the evolutionary process domain knowledge that is itself largely generated as part of the design process.

References

Beer, R. D. 1995. A dynamical systems perspective on agent-environment interaction. *Artificial Intelligence* 72:173–215.

Bigge, W. T. 2010. The programmable spring: Towards physical emulators of mechanical systems. Doctoral thesis, University of Sussex.

Brooks, R. A. 1991. Intelligence without representation. *Artificial Intelligence* 47:139–159.

Harvey, I. 1992. Species adaptation genetic algorithms: A basis for a continuing SAGA. In *Towards a Practice of Autonomous Systems: Proceedings of the 1st European Conference on Artificial Life*, ed. F. J. Varela and P. Bourgine, 346–354. Cambridge, MA: MIT Press.

Harvey, I. 2001. Artificial evolution: A continuing SAGA. In *Evolutionary Robotics: From Intelligent Robots to Artificial Life*, LNCS 2217, ed. T. Gomi, 94–109. Heidelberg: Springer-Verlag.

Harvey, I., E. Di Paolo, R. Wood, M. Quinn, and E. Tuci. 2005. Evolutionary robotics: A new scientific tool for studying cognition. *Artificial Life* 11 (1–2): 79–98.

Harvey, I., P. Husbands, and D. Cliff. 1993. Issues in evolutionary robotics. In *From Animals to Animats 2: Proceedings of the 2nd International Conference on Simulation of Adaptive Behaviour (SAB92)*, ed. J.-A. Meyer, H. Roitblat, and S. Wilson, 364–373. Cambridge, MA: MIT Press.

Husbands, P. 1994. Distributed coevolutionary genetic algorithms for multi-criteria and multi-constraint optimisation. In Evolutionary Computing, AISB Workshop Selected Papers, vol. 865 (LNCS), ed. T. Fogarty, 150–165. Heidelberg: Springer-Verlag.

Jakobi, N. 1998. Minimal simulations for evolutionary robotics. Doctoral thesis, University of Sussex.

Kodjabachian, J., and J.-A. Meyer. 1998. Evolution and development of neural networks controlling locomotion, gradient following and obstacle avoidance in artificial insects. *IEEE Transactions on Neural Networks* 9:796–812.

Manoonpong, P., T. Geng, T. Kulvicius, B. Porr, and F. Wörgötter. 2007. Adaptive, fast walking in a biped robot under neuronal control and learning. *PLoS Computational Biology* 3 (7): e134. doi:10.1371/journal.pcbi.

McGeer, T. 1990. Passive walking with knees. In *Proceedings of the IEEE Conference on Robotics and Automation*, vol. 2, 1640–1645. Piscataway, NJ: IEEE Press.

Pratt, J. and G. Pratt. 1999. Exploiting natural dynamics in the control of a 3d bipedal walking simulation. In *Proceedings of the International Conference on Climbing and Walking Robots (CLAWAR99)*, ed. D. Howard, G. S. Virk, and M. J. Randall, 797–807. New York: John Wiley and Sons Ltd.

Raibert, M. H. 1986. *Legged Robots That Balance*. Cambridge, MA: MIT Press.

Robinson, D. W., J. E. Pratt, D. J. Paluska, and G. A. Pratt. 1999. Series Elastic Actuator development for a biomimetic robot. In *Advanced Intelligent Mechatronics, 1999. Proceedings of the 1999 IEEE/ASME International Conference on Advanced Intelligent Mechatronics*, ed. K-M. Lee and B. Siciliano, 561–568. Piscataway, NJ: IEEE Press.

Vaughan, E. 2007. The evolution of the omni-directional bipedal robot. Doctoral thesis, University of Sussex.

Vaughan, E., E. Di Paolo, and I. Harvey. 2004a. The evolution of control and adaptation in a 3d powered passive dynamic walker. In *Proceedings of the Ninth International Conference on the Simulation and Synthesis of Living Systems (ALIFE '9 Boston)*, ed. J. B. Pollack, M. Bedau, P. Husbands, T. Ikegami, and R. A. Watson, 139–145. Cambridge, MA: MIT Press.

Vaughan, E., E. Di Paolo, and I. Harvey. 2004b. The tango of a load balancing biped. In *Proceedings of CLAWAR 2004, 7th International Conference on Climbing and Walking Robots*, ed. M. Armada and P. González de Santos, 813–824. Heidelberg: Springer.

12 Mindless Intelligence: Reflections on the Future of AI

Jordan B. Pollack

12.1 Introduction

Most of my thirty-three years of professional involvement in AI (artificial intelligence) have been focused on research far from its mainstream, not because of any antisocial tendencies on my part, but because of certain dilemmas inherent in the field. The first dilemma confronting AI is that both single-celled and multicelled animals survive and reproduce very well without any nervous system at all, and "lower animals," even insects, organize into thriving societies without any symbols, logic, or language, bee dancing and birdsong notwithstanding. These phenomena led me to delve into non-symbolic models and ask how complex hierarchal representations and sustained state-changing procedures might naturally emerge from iterative numeric systems such as associative or connectionist neural networks.

The second dilemma is that the kind of mind we in AI seek to discover, one that "runs" on the human brain yet might be portable to another universal machine, wouldn't even exist without having coevolved with the brain—a chicken-and-egg problem. So, while many of my connectionist colleagues migrated with U.S. National Institutes of Health funding into cognitive or computational neuroscience, trying to understand how the human brain works, I focused instead on what natural process could design and fabricate machinery as complex as the brain.

I ended up working closer to the field of artificial life, seeking to understand how evolution, a mindless iterative reproduction system, could eventually lead to machines whose complexity and reliability dwarfs the product of the largest teams of human engineers.

AI is now more than fifty years old, so I would like to reflect on what I feel has been its great mistake, and propose a corrective course going forward. But before analyzing this mistake, I want to say that AI is a great human endeavor with a colorful cast and many partial successes. It has provided frameworks for formally studying biological systems, animals, and humans and has spun out industries such as Lisp machines, expert systems, data mining, and even Internet search.

12.2   Don't Promise the Practically Impossible

We all agree on AI's fundamental hypothesis, that physical machines have the capacity for intelligence. Unfortunately, this hypothesis can neither be proven nor refuted scientifically, but realized only by demonstration. And until it has been convincingly demonstrated, it must remain in scientific limbo. Ordinary citizens and funding bureaucrats don't know whether AI is *tardy*, like mechanical flight, which emerged from limbo after several hundred years of failure, or *magical*, like ESP or the alchemists' quest to turn lead into gold. Perhaps there is even an *impossibility proof* waiting around the corner, as has put to rest quixotic notions such as time travel (Einstein) and perpetual motion (Ludwig Boltzmann). Who wants to fund a field that might be proven impossible tomorrow?

So AI, which represents one of the greatest intellectual and engineering challenges in human history—and should command the same fiscal resources as efforts to cure cancer or colonize Mars—is sometimes relegated to a laughingstock status, because we can't prevent bogus claims from cropping up in newspapers and books. We cannot seem to convince the public that humanoids and Terminators are just Hollywood special effects, as science-fictional as the little green men from Mars!

Still, some want to keep pursuing the same old AI goals: "What are the missing pieces necessary to achieving human-level common sense?" "Let's do a project to gain human-level performance in a (non-chess) domain." "We will build natural language software that's human-level in ability." "Soon computers will be fast enough to supply human-level intelligence to humanoid robots."

AI won't be a gift of more CPU time. If it were, we would have already glimpsed real AI on supercomputers or large clusters, yet nothing of the kind has occurred. We don't need faster chips to make robots smarter, since we can link a robot's body to its supercomputer brain over wireless broadband. As the joke goes, even if AI requires an infinite loop, it should run in only five seconds on a supercomputer.

The issue isn't the speed of running a mind-like program; it is the size and quality of the program itself. Because we routinely underestimate the complexity of evolved biological systems, and because Moore's law doesn't lead to a doubling of the quality of human-written software (Lanier 2000), the same old goals are red herrings that promise the practically impossible!

12.3   Take Mind Off Its Pedestal

AI's great mistake is its assumption that *human-level intelligence is the greatest intelligence that exists*, and thus, that our computational intelligences should operate "like" human cognition. Because of this mistake, most AI research has focused on "cognitive models" of intelligence, on programs that run like people think. But it turns out that we don't think the way we think we think!

The scientific evidence coming in all around us is clear: 'Symbolic Conscious Reasoning,' which is extracted through protocol analysis from serial verbal introspection, is a myth. From Michael Gazzaniga's famous split-brain experiments, where a patient associated a snow shovel with a chicken (Gazzaniga 1985), through Daniel Dennett's (1991) demolition of consciousness, through the unconscious intelligence described recently by Malcolm Gladwell (2005), it's entirely clear that the "symbolic mind" that AI has tried for more than fifty years to simulate is just a story we humans tell ourselves to predict and explain the unimaginably complex processes occurring in our evolved brains.

Because of this preoccupation with mimicking human-level intelligence, as a scientific field, AI has largely ignored or excluded the contributions of many alternative nonsymbolic mechanisms. Such mechanisms range from associative and matrix models of mathematical psychology, to Markovian models, to both game and decision theories, to early neural networks (the perceptron disaster), to simulations of evolution and organic self-organization. The early success of low-hanging symbolic fruit through Lisp programming led to the pursuit of the "mythical man module," a computer program that has the "look and feel" of human cognition yet is something more than an Eliza.

John Searle's "Chinese Room" argument (Searle 1980) is hateful because, in fact, he's correct. Neither the room nor the guy in it pushing symbols "understands" Chinese. But this isn't really a problem, because nobody actually "understands" Chinese! We only think we understand it. As anyone—even a native speaker—drives further down into an explanation of his or her knowledge or behavior, instead of gaining sharper insights (as we might expect in a reductionist physical science with a better microscope), the explanations get blurrier and blurrier.

By assuming that intelligences based on human-centric cognitive architectures such as grammars or production systems are the zenith, are the most powerful intelligences in the world, our field has made the same kind of embarrassing mistake as today's cryptocreationists, the proponents of Intelligent Design: by doubting that a mindless nonlinear iterative process such as evolution could be responsible for irreducible complexity in the designs of biological life forms, they hold that a superhuman, superintelligent being must have intervened.

AI also behaves as if human intelligence is next to godliness. Even the neural approach, more accepted today then ever, falls into the trap of trying to model human cognitive structures such as verb conjugation. Why is simulating the human mind more important than simulating cellular metabolisms, insect or animal intelligence, complex pattern formation, or distributed control of complex ecologies? It must be because, as a mirror of our own intelligence, the mindless iterative and numeric computing we scientifically uncover in nature doesn't compare to the perfectly logical indefatigable mind of Hollywood characters such as Mr. Spock and Commander Data, NP-completeness notwithstanding.

To repair this mistake and move forward as a scientific field, AI must recognize that many intelligent processes in nature perform more powerfully than human symbolic reasoning, even though they lack any of the mind-like mechanisms long believed necessary for human "competence." Once we recognize this and start to work out these scalable representations and algorithms without anthropomorphizing them, we should be able to produce the kind of results that will get our work funded to the level necessary for growth and deliver beneficial applications to society, without promising the intelligent English-speaking humanoid robot slaves and soldiers of science fiction.

12.4   Defining Mindless Intelligence

I define "mindless intelligence" as intelligent behavior ascribed (by an observer) to any process lacking a mind-brain. Suppose some black-box process (for example, mathematical, numerical, or mechanical) exhibits behavior that appears to require intelligence. However, when we scientifically study it, we find no Lisp interpreter, no symbols, no grammars, no logic or inference engine—in fact, we realize that it works without any of the accoutrements of cognition. We can say that this process is mindlessly intelligent.

Now we can begin to seriously study intelligent performance by

• feedback-driven systems such as thermostats and steam governors;
• pattern-action systems such as Eliza programs and immune systems;
• stability and hierarchy networks such as cellular metabolisms;
• societal assemblies such as insect and colonial life forms;
• utility-maximizing systems such as game and economic agents;
• exquisitely iterative systems such as evolution, fractals, and embryogenesis; and even
• mind-erasing collectives such as academic committees, crowds, and bureaucracies.

To give you a broader sense of the field, I'll briefly cover several kinds of natural processes that appear intelligent yet lack any cognitive apparatus. John Kolen and I showed how an iterated dynamical system could appear to generate a context-free or context-sensitive language, depending on the observer (Kolen and Pollack 1995). The dynamical system lacked any cognitive architecture for "generative capacity," which has been assumed by all natural language processing systems since Noam Chomsky.

Wherever we look in nature, we see amazingly complex processes to which we can ascribe intelligence, yet we observe symbolic cognition in only one place, and only there as a result of introspection. Many of these natural processes have been studied under the aegis of complex systems or have been given the prefix "self" or "auto." Because these systems have no mind, and thus no self, I've taken the liberty of replacing those prefixes with the new term "ectomental," which means "outside" (Greek) "of mind" (Latin).

12.5   Ectomental Organization

Evolution is the primary example of an intelligent designer who lacks a mind. There's no grammar, set of rules, library of CAD parts, or physics simulation. Simply put: a mindless reproductive system operates, transcription errors occur, and selection locks in a statistical advantage for the marginally better—or luckier—members of a population. And yet this iterative process has automatically designed machines of incredible beauty and complexity, objects that far surpass—in complexity and reliability—anything architects, engineers, novelists, venture capitalists, or teams of software programmers can achieve.

Human teams can build systems with only 10 million to 100 million unique moving parts before the entire structure collapses, yet biological forms can have 10 billion unique moving parts.

For the past decade and more, my lab's goal has been to understand how evolution can produce more complex designs than a human engineering team, while lacking human-level symbolic cognition. We've focused specifically on coevolutionary machine learning systems. While we haven't yet achieved a fully open-ended design process, we have

• shown coevolutionary systems that have surpassed human performance in sorting networks and cellular-automata optimization (Juille and Pollack 1999);
• developed theories such as Pareto coevolution (Ficici and Pollack 2001), emergent dimensionality (Bucci and Pollack 2003), and computational models of symbiogenesis (Watson and Pollack 2002); and
• revealed the possibility of motivating a community of learners (Sklar and Pollack 2000) to become their own Ideal Teachers (DeJong and Pollack 2004), resulting in novel educational software (Bader-Natal and Pollack 2005).

Perhaps our best-known research is on the coevolution of robot bodies and brains, known as the Genetically Organized Lifelike Electro-Mechanics, or GOLEM, project. This research resulted in three generations of self-designed systems that discovered irreducibly complex components and processes such as the cantilever, ratcheting, and kayaking (see figure 12.1).

Figure 12.1
Three generations of evolved robots: (a) Pablo Funes's evolution of Lego discovered the cantilever (Funes and Pollack 1998), (b) Hod Lipson's evolution of dynamic trusses invented the ratchet (Lipson and Pollack 2000), and (c) Gregory Hornby's evolution using L-systems to describe machines invented a kayaking motion (Hornby and Pollack 2002).

12.6 Ectomental Learning

One of the oldest AI paradigms is a self-learning or autodidactic system, a program that begins with a tabula rasa and, when dropped into an environment, gets better and better over time. Perhaps the best example of such a system is Gerald Tesauro's TD-Gammon (Tesauro 1992). He started with essentially a random neural network that could return a value for any backgammon position. Rather than training the network against an encyclopedia of human expert games, he essentially trained it against itself. After about a month of computer time on an IBM supercomputer, with the weights adjusted as a result of each game, his network, with further refinements, became one of the best players in the world.

Humans can verbalize backgammon strategies. We consider only a few plausible moves and then estimate whether one move is better or worse than another on the basis of strategic goals from models of the game (running, blocking, back-game), using all kinds of approximate and exact calculations about probability. I was a professional-level backgammon player in 1975 and felt that there were about seven different human-player "types" who, at the top of their game, achieved a rock-scissors-paper parity.

On the other hand, TD-Gammon is a mindless intelligence that dominates all human players. It uses a function to estimate values and uses a one- or two-ply look-ahead with a greedy selector to make a move. It has no logic or symbols, no strategy that looks far ahead or back in time, and no language component to discuss its strategy. Yet it's stronger than any rule-based strategy.

My lab had worked on self-learning for tic-tac-toe (Angeline and Pollack 1993), and we became interested in understanding why TD-Gammon worked. We were able to replicate the Tesauro effect using simple hill climbing (Pollack and Blair 1998), which led to the question of why coevolutionary self-learning worked so well for backgammon. Game theorists such as Richard Bellman recognized many years ago why a purely numeric backgammon player works better than a logical game (Bellman 1957). He proved the existence of a value table for optimal sequential choice in Markovian games, where opponents can choose strategies yet are buffeted by random elements such as dice. Moreover, iterated approximation of the value table, through a single-ply expectimax look-ahead, leads to its convergence. So, an optimal value table combined with a one-ply greedy choice leads to the strongest-possible player.

In order to study the success of learning backgammon, I recently invented Nannon®, the smallest version of backgammon that maintains its core behaviors, using only six points, three checkers, and one die per side. There are only 2,530 different board positions, and the value table converges in 15 sweeps to an error of 10^{-7} (Pollack 2005). While the full game of backgammon is much larger than Nannon, so a table can't be stored, Tesauro's choice of input representation and network size from earlier experiments led to a fortuitous convergence between TD reinforcement learning and Bellman's earlier mathematical work. Perhaps many mindlessly intelligent processes in nature are similar instances of mathematical ideals that can lead to convergence, complexity, and optimal performance in the limit.

12.7   Ectomental Repair

A marvelous characteristic of natural systems is that they can heal, or self-repair. A naïve computerized view would be to envision the algorithmic equivalent of a team of repairpersons who, under centralized supervision, consult a system model and are then deployed to a disturbance's site to apply cognition, logic, and spare parts to return the system to model behavior. However, imagining a system that contains a deployable model of itself can lead to logical conundrums (Minsky 1965).

How might we understand self-repair in natural systems? In artificial-life research on "algorithmic chemistry," Walter Fontana and Leo Buss described systems of simple lambda calculus programs that consume and produce each other, forming a metabolism (Fontana and Buss 1994). When such an artificial-chemistry network had a steady-state dynamic, perturbations would return to the same attractor, like the memories in a Hopfield network.

Is the *Bauplan* of an animal a similar attractor, which the myriad of microscopic mindless actions can't help but keep returning to? In other words, the answer to self-repair is that there's no blueprint or explicit diagram; there's just a framework and a set of parameters that mathematically define a complex attractor. Mindless and far-flung distributed operations can't help themselves; they must gravitate toward it.

Such dynamical systems with complex attractors driven by parameters are well known. One example is the Mandelbrot set, a truly exquisite iteration where the parameters define a window and each pixel computes its own color. Another example is *iterated function systems* (IFS), a union of a set of contractive maps that Michael Barnsley proved has a single fractal limit attractor akin to Cantor dust (Barnsley 1988).

Barnsley showed, much analogous to Bellman's proof, that some nonlinear iterative processes, despite having many adjustable parameters, have a single, yet complicated, limit, defined by the interaction of the parameters and rules. Simply put, an IFS fractal attractor is like repeatedly copying an image with a special copying machine that makes multiple shrunken and transformed copies of the input page (see figure 12.2). All nonblank starting pages, from a speck of dust to a piece of black construction paper, end up converging to the same attractor in the limit.

I came across IFSs while working to understand the relationship between recurrent neural networks and finite-state machines. As the result of trying to learn a language, a recurrent network generated an infinite-state machine with the states located on a fractal attractor (Pollack 1991). Subsequent research used these structures for memory and hierarchal representations (Levy, Melnik, and Pollack 2000).

a   b

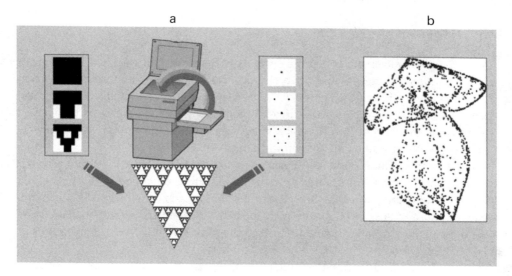

Figure 12.2

(a) An iterated-function-systems fractal is like a feedback loop on a copy machine that makes more than one reduced copy of an image, resulting in the same limit for a speck of dust or a full page of ink. (b) The IFS theory explained the "strange automata" that emerged when recurrent neural networks were trained to recognize languages.

The mindless intelligence of self-defining and self-repairing, or autopoetic (Varela, Maturana, and Uribe 1974), biological forms is a big leap from Fontana's chemistries and Barnsley's fractals. Yet I am certain that biological form will one day be scientifically explained as an attractor that changes its parameters over time while it's constantly and mindlessly repaired by distributed processing at a microscopic level.

12.8 Ectomental Assembly

Fetal development, or embryogenesis, is perhaps the perfect place to recognize the profound scale of complex behavior achievable by mindless intelligence.

Herb Simon introduced Tempus and Hora as two different kinds of watchmakers who suffer from interruptions: one uses modular construction; the other works with basic parts (Simon 1969). Richard Dawkins introduced the idea of the Blind Watchmaker (Dawkins 1986). Both researchers comfortably anthropomorphized what is a mindless assembly process.

Every assembly factory depends critically on human minds both as labor as well as supervision to monitor, correct, and repair ongoing processes. Yet a developing fertilized egg is also an assembly factory, without any human supervisors or any brain, which produces an exquisite, custom product with ten billion moving parts in only nine months! Where's the mind inside the fertilized egg? Even Intelligent Design proponents might be hard pressed to defend the existence of an omniscient "Intelligent Factory Foreman" who supervises every embryo developing in the world simultaneously, deciding which creatures live or die.

Other than basic work on pattern formation, related to work by, for example, Alan Turing and Stephen Wolfram, we have a long way to go in understanding the mindless intelligence in a process that could self-assemble into a biological form with billions of parts. My lab is working on replacing the idea of a perfect robotic factory with evolutionary processes that must evolve both form and formation and overcome noise and error in physical assembly (Rieffel and Pollack 2005). One of the more interesting threads is the relationship between robotic assembly with errors and noise, and the kinds of tasks that Bellman proved could iteratively converge to optimal (Viswanathan and Pollack 2005). This might provide a self-construction theory involving not a blind watchmaker but a *blind chess master* who continuously optimizes assembly processes to maximize its own chances for successful reproduction.

12.9 Ectomental Reproduction

Another great mystery of nature is complex self-reproduction. Shy of a magical reverse-engineering theory (which would let us genetically engineer flying horses), we have little or no grasp on the algorithmic processes involved in the major transition from single cells reproducing, through colonialization, to multicellular creatures with differentiated tissues and functions.

I think it's another case of dramatically underestimating the amount of intelligence in a seemingly obvious natural process. We have many simple examples of reproduction in software, from straight data copying to self-reproducing code as shown by evaluating this ditty in Common Lisp:

```
((LAMBDA (X) (LIST X (LIST 'QUOTE X)))
'(LAMBDA (X) (LIST X (LIST 'QUOTE X))))
```

Following John Von Neumann's challenge of finding self-reproduction in cellular automata, Christopher Langton helped birth the field of artificial life with his more elegant automata (Langton 1984), and Jason Lohn and James Reggia showed how easy it is to discover the rules for such automata (Lohn and Reggia 1997).Yet so far, all our computing reproducing systems, including Tom Ray's Tierra (Ray 1991) and Hod Lipson's cubes (Zykov et al. 2005), are very simple. I'm hopeful that evolutionary search for more complex reproductive forms holds some hope for understanding how a mindless reproductive process can become more capable over time to sustain complexity in the design of reproducible machinery.

12.10   Ectomental Recognition, Control, and Regulation

Obviously, intelligence arises outside the mental sphere in so many other places in nature that I can't list them all.

The immune system is an ectomental chemical recognition system that filters and separates millions of chemicals along the me/not-me boundary, without a central database listing which compounds are in or out. Self-control of physical movement, of individuals and groups, is often mindless. This isn't only because time constraints push nervous-system controls to the edge but also because it's hard to find a valuable use for cognitive symbols inside mainly numeric models such as pattern generators and feedback loops.

Finally, the zenith of self-regulation is probably the planet itself. Similar to Adam Smith's "invisible hand" idea that markets are mindlessly intelligent regulators and allocators of goods and services, the Gaia hypothesis proposes that the whole biosphere operates so as to maintain the right conditions for life as we know it (Lovelock 1979). A trivial and kooky interpretation is that Gaia is a goddess with a mind of her own, complete with symbols, logic, and language, so she might talk to us one day through a burning bush or a statue of her likeness. A deeper interpretation is to recognize that the algorithmic complexity of balancing resources, encouraging growth, and managing the network of species to maintain the "sweet spot" for life is a huge job requiring such intelligence that we better not entrust it to any elected human officials!

Under the mindless-intelligence viewpoint, both evolution itself and the global-regulation system known as Gaia are intelligent beyond and outside the mental framework based on the symbol manipulation that AI has chosen as its focus.

I'm neither alone nor unique in wishing for a stronger scientific basis for the field. These comments certainly hearken back to many earlier calls (Brooks 1991). Much of the world has changed in the last decade. For example, after so many years of chasing generative linguistics' focus on parsing and syntax, the main thrust of both natural language processing and speech recognition has been to drive mindless statistical responses from large corpora rather than to establish carefully wrought rules and features. Intelligent-control research is also moving in a mindless way, from robotics that use shaky logical algorithms to more mathematically sophisticated nonlinear control systems (Zhao 1994). Much cognitive modeling research takes seriously the idea that algorithms should be not only cognitively plausible but also neurally plausible. Finally, machine learning research has progressed from its early efforts at matching human learning curves, to building strong algorithms for extracting knowledge from large statistical sources.

Yet these fields often must defend themselves from the charge that they aren't really AI. A few years ago George Dyson visited Google and wrote that he has long considered that when "real" AI arrives on the scene, it will be surrounded by "a circle of cheerful, contented, intellectually and physically well-nourished people" (Dyson 2005). Certainly Google is based on a very large database and uses statistical machine learning techniques to choose which keywords are important in different contexts. Does Google software have any of the cognitive aspects that AI has studied for many years? The mindless market doesn't care.

As we've seen, mindless intelligence abounds in nature, through processes that channel mathematical ideals into physical processes that can appear optimally designed yet arise through and operate via exquisite iteration.

The hypothesis for how intelligence arises in nature is that dynamical processes, driven by accumulated data gathered through iterated and often random-seeming processes, can become more intelligent than a smart adult human, yet continue to operate on principles that don't rely on symbols and logical reasoning. The proof lies not only in Markovian situations where a greedy sequential-choice algorithm driven by values converged under Bellman's equation, but also in the reliability, complexity, and low cost of biologically produced machines.

Because our minds aren't what they seem, symbolic explanations of our behavior that were extracted from protocol analysis and conscious introspection are misleading at best and complete fabrications at worst. Most of what our brains are doing involves mindless chemical activity not even distinguishable from digestion of the food in the Chinese Room.

I don't mean to imply that human cognition isn't worth studying. I just want to reiterate that cognitive reporting is an always-incomplete story, a simplified verbalization of a partial insight of the working patterns of our brains. And brains aren't instruction set computers; they're complicated biological networks with all kinds of feedback at all levels, like metabolisms, gene regulatory networks, and immune systems. The software and systems that emerge from and control these networks, like evolution, embryological-development protocols, Gaian ecological regulation, or mind, will be much harder to reverse engineer than the artifacts of human engineering culture.

Emphatically then, as AI arises, it won't be organized like a good computer program, it won't speak English, and it certainly won't act like a humanoid robot from a science fiction movie. 'Symbolic Mind' is a self-aggrandized fiction told to make sense of a few pounds of mindlessly intelligent meat. It's time we wean ourselves from the fiction and start working on the science.

Acknowledgments

Thanks to my Ph.D. students, past and present, for their collaborations, and to Carl Feynman for help coining a word.

Notes

This chapter is based on "Mindless Intelligence" by Jordan Pollack, which appeared in *IEEE Intelligent Systems* 21(3): 50–56. © 2006 IEEE. Reproduced with permission.

References

Angeline, P., and J. Pollack. 1993. Competitive environments evolve better solutions to complex problems. In *Proceedings of the 5th International Conference on Genetic Algorithms (ICGA '93)*, ed. S. Forrest, 264–270. San Francisco, CA: Morgan Kaufmann.

Bader-Natal, A., and J. Pollack. 2005. Motivating appropriate challenges in a reciprocal tutoring system. In *Proceedings of the Fifth International Conference on Artificial Intelligence in Education 2005 (AIED '05)*, ed. C-K. Looi, G. I. McCalla, B. Bredeweg, and J. Breuker, 49–56. Amsterdam: IOS Press.

Barnsley, M. 1998. *Fractals Everywhere*. Boston: Academic Press.

Bellman, R. 1957. *Dynamic Programming*. Princeton, NJ: Princeton University Press.

Brooks, R. 1991. Intelligence without representation. *Artificial Intelligence* 47 (1–3): 139–159.

Bucci, A., and J. Pollack. 2003. Focusing versus intransitivity: Geometrical aspects of coevolution. In *Proceedings of the 2003 Genetic and Evolutionary Computation Conference*, ed. E. Cantú-Paz, J. A. Foster, K. Deb, L. D. Davis, R. Roy, U-M. O'Reilly, H-G. Beyer, R. Standish, G. Kendall, S. Wilson, M. Harman, J. Wegener, D. Dasgupta, M. A. Potter, A. C. Schultz, K. A. Dowsland, N. Jonoska, and J. Miller, LNCS 2723, 250–261. Berlin: Springer.

Dawkins, R. 1986. *The Blind Watchmaker: Why the Evidence of Evolution Reveals a Universe without Design*. New York: W. W. Norton and Co.

De Jong, E., and J. Pollack. 2004. Ideal evaluation from coevolution. *Evolutionary Computation* 12 (2): 159–192.

Dennett, D. 1991. *Consciousness Explained*. Boston: Little, Brown and Co.

Dyson, G. 2005. Turing's cathedral, *Edge*, October 24. www.edge.org/3rd_culture/dyson05/dyson05_index.html (accessed June 11, 2013).

Ficici, S., and J. Pollack. 2001. Pareto optimality in coevolutionary learning. In *Advances in Artificial Life: 6th European Conference (ECAL 2001)*, LNAI 2159, ed. J. Kelemen and P. Sosik, 316–325. Berlin: Springer.

Fontana, W., and L. Buss. 1994. The arrival of the fittest: Toward a theory of biological organization. *Bulletin of Mathematical Biology* 56:1–64.

Funes, P., and J. Pollack. 1998. Evolutionary body building: Adaptive physical designs for robots. *Artificial Life* 4 (4): 337–357.

Gazzaniga, M. 1985. *The Social Brain: Discovering the Networks of the Mind*. New York: Basic Books.

Gladwell, M. 2005. *Blink: The Power of Thinking without Thinking*. New York: Little, Brown and Co.

Hornby, G., and J. Pollack. 2002. Creating high-level components with generative representation for body-brain evolution. *Artificial Life* 8 (3): 223–246.

Juille, H., and J. Pollack. 1999. Coevolutionary learning and the design of complex systems. *Advances in Complex Systems* 2 (4): 371–393.

Kolen, J., and J. Pollack. 1995. The observers' paradox: Apparent computational complexity in physical systems. *Journal of Experimental & Theoretical Artificial Intelligence* 7 (3): 253–269.

Langton, C. 1984. Self-reproduction in cellular automata. *Physica D. Nonlinear Phenomena* 10 (1–2): 135–144.

Lanier, J. 2000. One half of a manifesto, *Wired*, December. www.wired.com/wired/archive/8.12/lanier.html (accessed June 11, 2013).

Levy, S., O. Melnik, and J. Pollack. 2000. Infinite RAAM: A principled connectionist basis for grammatical competence. In *Proceedings of the 22nd Annual Conference of the Cognitive Science Society,* ed. L. R. Gleitman and A. K. Joshi, 298–303. http://csjarchive.cogsci.rpi.edu/Proceedings/2000/COGSCI00.pdf (accessed June 11, 2013).

Lipson, H., and J. Pollack. 2000. Automatic design and manufacture of robotic life-forms. *Nature* 406 (6799): 974–978.

Lohn, J., and J. Reggia. 1997. Automatic discovery of self-replicating structures in cellular automata. *IEEE Transactions on Evolutionary Computation* 1 (3): 165–178.

Lovelock, J. 1979. *Gaia: A New Look at Life on Earth.* Oxford: Oxford University Press.

Minsky, M. 1965. Matter, mind, and models. In *Proceedings of the IFIP Congress,* ed. W. A. Kalenich, 45–49. http://groups.csail.mit.edu/medg/people/doyle/gallery/minsky/mmm.html (accessed June 11, 2013).

Pollack, J. 1991. The induction of dynamical recognizers. *Machine Learning* 7 (2–3): 227–252.

Pollack, J. 2005. Nannon: A nano backgammon for machine learning research. In *Proceedings of the 2005 International Conference on Computational Intelligence in Games,* ed. G. Kendall and S. Lucas, 277–284. http://cswww.essex.ac.uk/cig/2005/papers/proceedings.pdf (accessed June 11, 2013).

Pollack, J., and A. Blair. 1998. Co-evolution in the successful learning of backgammon strategy. *Machine Learning* 32 (3): 225–240.

Ray, T. 1991. An approach to the synthesis of life. In *Artificial Life II,* ed. C. Langton et al., 371–408. Redwood City, CA: Addison-Wesley.

Rieffel, J., and J. Pollack. 2005. Evolving assembly plans for fully automated design and assembly. In *Proceedings of the NASA/DoD Conference on Evolvable Hardware (EH '05),* ed. J. Lohn, D. Gwaltney, G. Hornby, R. Zebulum, D. Keymeulen, and A. Stoica, 165–170. Los Alamitos, CA: IEEE Press.

Searle, J. 1980. Minds, brains, and programs. *Behavioral and Brain Sciences* 3 (3): 417–457.

Simon, H. 1969. *The Sciences of the Artificial.* Cambridge, MA: MIT Press.

Sklar, E., and J. Pollack. 2000. A framework for enabling an internet learning community. *Journal of Educational Technology & Society* 3 (3): 393–408.

Tesauro, G. 1992. Temporal difference learning of backgammon strategy. In *Proceedings of the International Conference on Machine Learning (ICML '92),* 451–457. San Francisco, CA: Morgan Kaufmann.

Varela, F., H. Maturana, and R. Uribe. 1974. Autopoiesis: The organization of living systems, its characterization and a model. *Bio Systems* 5 (4): 187–196.

Viswanathan, S., and J. Pollack. 2005. On the robustness achievable with stochastic development processes. In *Proceedings of the NASA/DoD Conference on Evolvable Hardware (EH '05),* ed. J. Lohn,

D. Gwaltney, G. Hornby, R. Zebulum, D. Keymeulen, and A. Stoica, 34–39. Los Alamitos, CA: IEEE Press.

Watson, R., and J. Pollack. 2002. A computational model of symbiotic composition in evolutionary transitions. *Bio Systems* 69 (2–3): 187–209.

Zhao, F. 1994. Extracting and representing qualitative behaviors of complex systems in phase spaces. *Artificial Intelligence* 69 (1–2): 51–92.

Zykov, V., E. Mytilinaios, B. Adams, and H. Lipson. 2005. Self-reproducing machines. *Nature* 435 (7038): 163–164.

Contributors

Christos Ampatzis
European Space Agency, Noordwijk, NL

Randall D. Beer
Indiana University, USA

Josh Bongard
Vermont University, USA

Joachim de Greeff
Institute of Cognitive Science and Technology, CNR, Italy

Ezequiel A. Di Paolo
Ikerbasque, University of the Basque Country

Marco Dorigo
Université Libre de Bruxelles, Belgium

Dario Floreano
EPFL, Switzerland

Inman Harvey
University of Sussex, UK

Sabine Hauert
Massachusetts Institute of Technology, USA

Phil Husbands
University of Sussex, UK

Laurent Keller
University of Lausanne, Switzerland

Michail Maniadakis
Computational Vision and Robotics Laboratory, FORTH, Greece

Orazio Miglino
University of Naples, Italy

Sara Mitri
University of Oxford, UK

Renan C. Moioli
Edmond and Lily Safra International Institute for Neuroscience of Natal (ELS-IINN), Brazil

Stefano Nolfi
Institute of Cognitive Science and Technology, CNR, Italy

Michael O'Shea
University of Sussex, UK

Rainer W. Paine
NINDS, National Institutes of Health (NIH), USA

Andy Philippides
University of Sussex, UK

Jordan B. Pollack
Brandeis University, USA

Michela Ponticorvo
University of Naples, Italy

Yoonsik Shim
University of Sussex, UK

Jun Tani
RIKEN Brain Science Institute, Japan

Vito Trianni
ISTC-CNR, Italy

Elio Tuci
Aberystwyth University, UK

Patricia A. Vargas
Heriot-Watt University, UK

Eric D. Vaughan
University of Sussex, UK

Index

Intelligent Robotics and Autonomous Agents
Edited by Ronald C. Arkin